Enquiring Minds AND Space Aliens

Also by Walter M. Brasch

A Comprehensive Annotated Bibliography of American Black English
(with Ila Wales)

Black English and the Mass Media

Columbia County Place-Names

Cartoon Monickers: An Insight Into the Animation Industry

The Press and the State: Sociohistorical and Contemporary Issues
(with Dana R. Ulloth)

Forerunners of Revolution: Muckrakers and the American Social Conscience

A ZIM Self-Portrait

With Just Cause: Unionization of the American Journalist

Betrayed: Death of an American Newspaper

Joel Chandler Harris, Uncle Remus, and the American Social Conscience
(forthcoming)

Enquiring Minds AND Space Aliens

Wandering Through the Mass Media and Popular Culture

by Walt Brasch

Mayfly Productions
Elmwood, Illinois 61529

Publisher: Bill Knight
Typesetting and cover design: Mark Steinruck
Sculpture: Auguste Rodin

Mayfly Productions
P.O. Box 380
Elmwood, Illinois 61529

Library of Congress Catalog Card Number: 95-81345
International Standard Book Number: 0-9624613-4-2

This book is printed on acid-free paper

PRINTED IN THE UNITED STATES OF AMERICA

Dedication . . .

. . . to Rosemary Renn Brasch who provided inspiration for many of the columns, and encouragement for all of them. Her ideas and suggestions are always good, her social conscience a model for others to emulate.

. . . And, as always, to my parents Milton and Helen Haskin Brasch who provided the inspiration to challenge injustice and stupidity while still being a part of the greater community of mankind.

Acknowledgements

During the past two years, I have been especially fortunate to have three excellent and enthusiastic assistants who have done everything from filing routine correspondence to listening to my gripes. Words and a pittance of a salary aren't enough to repay them for their time, energy, and loyalty. But, let it be known that Jennifer Boscia, Meka Eyerly, and Mark Steinruck have all been a part of the column in ways that are not reflected in words.

Other assistance was provided by Renee Boyer, Linda Haines, Tony Ianiero, Markland Lloyd, and Jeremy Powlus.

Every writer also needs an editor, although we often wonder why. With more than four dozen editors over the past three years, the frustration could have become immense. Fortunately, I have found editors who have usually provided valuable advice and friendship, even if now and then they have to edit or cut a column. I am sure that many of the columns may have caused them to hide behind bunkers, awaiting the volley of controversy that results. I am especially pleased to recognize and thank editors Nancy Acitelli, Wayne Bauer, Jacob Betz, Jim Burchik, Shawnee Culbertson, Jim Curtis, Vivian Daily, Sharon Falkowitz, Joseph X. Flannery, Jim Gallagher, David Gilmartin, Ann Marie Gonsalves, Andrew Gross, Bekki Guilyard, Bill Kohler, Mike Lion, Leland B. Mather, Holly Matthews, Diane Mazonis, Marie McCandless, Wiley McKellar, Pat McKenna, Dave Monaghan, John Moore, James Oliver, Mitchell Olszak, Harold Prentiss, Michael Regan, Ray Saul, Charles Schenk, Ed Schreppel, Linda Seligson, Jim Seybert, Joseph P. Shaw, Joe Shea, Dick Shearer, Joe Sukle, Eileen Winter, and Jack Yoset.

Two people who I never met helped shape my thought process. Horace Greeley (1811-1872), editor/publisher of the *New York Tribune* and one of the most influential people in the history of our country, showed just how good journalism can be when it's done well. Art Buchwald (1925-), the nation's finest contemporary satirist, follows in the tradition of Horace, Juvenal, Rabelais, Cervantes, Moliere, Swift, Addison, Twain, William Gilbert, Sinclair Lewis, and James Thurber. They have shown just how powerful humor and satire can be in making people think and better understand their lives and their society.

Finally, but most important, a special thanks to all my readers, especially those who took the time to write or to call, even if they disagreed with my views.

Contents

Dedication v
Acknowledgements vii
Contents ix
Contents by Medium xii
Contents by Issue xvi
Introduction xxv

Enquiring Minds and Space Aliens 3
The Case of the Missing Popularity 6
Spinning Their Way to Success 8
Promises, Promises 10
Dishonesty: The Path to Public Confidence 12
Electing the News 14
A Columnist's Best Friend 16
The Joy of Sax 18
President Sneezes; Film at 11 19
A Dam(n) for the Home Folks 21
Long Island Lolita Meets the Journalistic Lowlife 23
The Only Alternative for KFAD 25
Blood on Their Lenses 27
A Major Conflict of Interest 29
The Intruder 30
TV News 101: Intro to Makeup 33
Politically Incorrect Weather 35
Reel Violence 37
An American Triangle: Violence, Media, and Individual Responsibility 39
Escalating an American Paranoia 42
Labels of Violence 45
Compliments of a Thief 47
Stripping Off Their Royalties 49
The Hustle 51
The Write Stuff 53

Rerunning America 55
'Local Reporter Cops Lexicon Story' 57
Linguistic Larceny 59
Editing the 12 Commandments 63
All in Vein 64
For Sale: Justice—O.J. Style 66
Dysfunctional Thoughts 68
Playing the University Grantsmanship Game 70
The $6 Million Journalist 72
Stealing the First Amendment 73
Creating Historical Fiction 75
The Press Meets the Afo-a-Kom 77
The Case for Sensationalism 80
Make Mine Media Rare 82
An Obscene Story 84
Of Matrons and Movies 86
Wonderings of an Idle Mind 87
America's Ding-a-Lings 90
The Beeper Cacophony 92
The Impersonal Society 94
Curbing the Paperless Explosion 96
Don't Despair, GM; Help is on the Way 98
Mixing a Bitter Pill for America's Drug Companies 100
Selling Out America by the Yard 102
The Day the Circus Came to Town 103
DATELINE: London Bridge, Arizona 105
Disneyland: Where Fantasies are Real 109
For Casinos, 'Six-Hour Visitors' Make a Full House 114
Conventional Fund Spending 117
The Monday Night Football Game 119
Sammy the Hustler 123
Three Nights a Week 126
Sounds of the Concrete City 128
Voices of America 130
A History of Fear 136
'And Now a Word From Your Local Sponsor' 140
War in the Gulf: Lessons of an Obsession 142
A Medal for the Army 146

Chicken Advertising 148
Kernels of Truth 150
Tanness, Anyone? 152
Cramping Up for Health 153
Death by Healthy Doses 155
The Stupid Season 157
All the Beautiful People 159
There She Is, Miss-representing America 162
Jeanetics and the Cool 'Wannabe' 165
Sex and the Single Beer Can 168
Facing a McBlimp Attack 173
Downsizing Baseball's Problems 175
Replacing Baseball 177
Traded for Two Rookies, An Editorial Clerk, and a Future Draft Choice 178
Laying Off Marshbaum 180
Patriotic Unemployment 182
An Unequal Competition 184
An Hour a Day 186
'We're Management: We Don't Have to Tell You Anything' 188
Withdrawing From an O.J. Overdose 192

Alternate Contents
(by Medium)

Books

Compliments of a Thief 47
Creating Historical Fiction 75
Stealing the First Amendment 73
Stripping Off Their Royalties 49

Film

Kernels of Truth 150
Of Matrons and Movies 86
The Hustle 51

Magazines

All the Beautiful People 159
Escalating an American Paranoia 42
Jeanetics and the Cool 'Wannabe' 165
The Press Meets the Afo-a-Kom 77

Mass Entertainment
(including music, recreation, health, and sports)

All the Beautiful People 159
For Casinos, 'Six-Hour Visitors' Make a Full House 114
Conventional Fund Spending 117
Cramping Up for Health 153
DATELINE: London Bridge 105
Death by Healthy Doses 155
Disneyland: Where Dreams Are Real 109
Downsizing Baseball's Problems 175
Dysfunctional Thoughts 68
Facing a McBlimp Attack 173
Replacing Baseball 177

Sammy the Hustler 123
Sex and the Single Beer Can 168
The Joy of Sax 18
The Monday Night Football Game 119
There She Is, Miss-representing America 162
Of Matrons and Movies 86
Sounds of the Concrete City 128
Stripping Off Their Royalties 49
Tanness, Anyone? 152
The Day the Circus Came to Town 103
The Hustle 51
The Stupid Season 157
Three Nights a Week 126
Voices of America 130

Miscellaneous

An Unequal Competition 184
Promises, Promises 10
Selling Out America by the Yard 102
The $6 Million Journalist 72
Wonderings of an Idle Mind 87

Newspapers

An Hour a Day 186
Curbing the Paperless Explosion 96
Dishonesty: The Path to Public Confidence 12
Editing the 12 Commandments 63
Electing the News 14
Enquiring Minds and Space Aliens 3
For Sale: Justice—O. J. Style 66
Laying Off Marshbaum 180
'Local Reporter Cops Lexicon Story' 57
Long Island Lolita Meets the Journalistic Lowlife 23
Make Mine Media Rare 82
President Sneezes; Film at 11 19
Stealing the First Amendment 73
The Case for Sensationalism 80
The Press Meets the Afo-a-Kom 77
Traded for Two Rookies, an Editorial Clerk, and a Future Draft 178

War in the Gulf:
Lessons of an Obsession 142
'We're Management:
We Don't Have to Tell You Anything' 188

Non-Specified Media

A Columnist's Best Friend 16
A History of Fear 136
A Medal for the Army 146
All in Vein 64
An Obscene Story 84
Chicken Advertising 148
Don't Despair, GM; Help is on the Way 98
Linguistic Larcency 59
Mixing a Bitter Pill for America's
Drug Companies 100
Patriotic Unemployment 182
Spinning Their Way to Success 8
The Case of the Missing Popularity 6
The Write Stuff 53

Personal Media

America's Ding-a-Lings 90
The Beeper Cacophany 92
The Impersonal Society 94

Point-of-Purchase

Sex and the Single Beer Can 168

Radio

For Sale: Justice—O. J. Style 66
'Local Reporter Cops Lexicon Story' 57
Make Mine Media Rare 82
Politically Incorrect Weather 35
President Sneezes; Film at 11 19
Reel Violence 37
The Press Meets the Afo-a-Kom 77
War in the Gulf:
Lessons of an Obsession 142
Wonderings of an Idle Mind 87

Television

A Dam(n) for the Home Folk	21
A Major Conflict of Interest	29
An American Triangle: Media, Violence, and Individual Responsibility'	39
And Now a Word From Your Local Sponsor'	140
Blood on Their Lenses	27
Dysfunctional Thoughts	68
For Sale: Justice—O. J. Style	66
Jeanetics and the Cool 'Wannabe'	165
Labels of Violence	45
Laying Off Marshbaum	180
'Local Reporter Cops Lexicon Story'	57
Long Island Lolita Meets the Journalistic Lowlife	23
Make Mine Media Rare	82
Playing the University Grantsmanship Game	70
President Sneezes; Film at 11	19
Promises, Promises	10
Rerunning America	55
The Intruder	30
The Monday Night Football Game	119
The Only Alternative for KFAD	25
The Press Meets the Afo-a-Kom	77
TV News 101: Intro to Makeup	33
War in the Gulf: Lessons of an Obsession	142
Withdrawing From an O.J. Overdose	192

Alternate Contents
(Contemporary Social Issues as Reflected in the Media)

Advertising

A Medal for the Army 146
'And Now a Word From Your Local Sponsor' 140
Chicken Advertising 148
Cramping Up for Health 153
Curbing the Paperless Explosion 96
Jeanetics and the Cool 'Wannabe' 165
Labels of Violence 45
Mixing a Bitter Pill for America's Drug Companies 100
Promises, Promises 10
Sex and the Single Beer Can 168

Business

An Hour a Day 186
'And Now a Word From Your Local Sponsor' 140
America's Ding-a-Lings 90
Conventional Fund Spending 117
Curbing the Paperless Explosion 96
Disneyland: Where Dreams Are Real 109
Don't Despiar, GM; Help Is On the Way 98
Downsizing Baseball's Problems 175
For Casinos, 'Six-Hour Visitors' Make a Full House 114
Jeanetics and the Cool 'Wannabe' 165
Laying Off Marshbaum 180
Mixing a Bitter Pill for America's Drug Companies 100
Patriotic Unemployment 182
Spinning Their Way to Success 8
Stripping Off Their Royalties 49

The Case of the Missing Popularity 6
The Day the Circus Came to Town 103

Censorship

An American Triangle; Media, Violence, and Individual Responsibility 39
An Obscene Story 84
Creating Historical Fiction 75
Labels of Violence 45
Make Mine Media Rare 82
Of Matrons and Movies 86
Stealing the First Amendment 73
War in the Gulf: Lessons of an Obsession 142

Checkbook Journalism

For Sale: Justice—O. J. Style 66
Long Island Lolita Meets the Journalistic Lowlife 23
Make Mine Media Rare 82
Wonderings of an Idle Mind 87

Crime and Violence

An American Triangle: Media, Violence, and Individual Responsibility 39
And Now a Word From Your Local Sponsor 140
Blood on Their Lenses 27
Escalating an American Paranoia 42
For Sale: Justice—O. J. Style 66
Labels of Violence 45
Make Mine Media Rare 82
Reel Violence 37
Rerunning America 55
Stripping Off Their Royalties 49
The Case for Sensationalism 80
TV News 101: Intro to Makeup 32
Withdrawing From an O.J. Overdose 192

Diversity

All the Beautiful People 159

An Hour a Day 186
Enquiring Minds and Space Aliens 3
Laying Off Marshbaum 180
There She Is, Miss-representing America 162
The Press Meets the Afo-a-Kom 77

Economics

America's Ding-a-Lings 90
Curbing the Paperless Explosion 96
Don't Despair, GM; Help Is On The Way 98
Dysfunctional Thoughts 68
Editing the 12 Commandments 63
Electing the News 14
For Casinos, 'Six-Hour Visitors' Make a Full House 114
Laying Off Marshbaum 180
Patriotic Unemployment 182
Politically Incorrect Weather 35
Replacing Baseball 177
The Day the Circus Came to Town 103
The $6 Million Social Journalist 72
The Stupid Season 157
Traded for Two Rookies, an Editorial Clerk, and a Future Draft Choice 178
'We're Management: We Don't Have to Tell You Anything' 188

Education

An Unequal Competition 184
Creating Historical Fiction 75
Playing the University Grantsmanship Game 70
The Stupid Season 157
The Write Stuff 53

Environment

Traded for Two Rookies, Editorial Clerk, and a Future Draft Choice 178

Ethics

A Dam(n) for the Home Folk 21
A Major Conflict of Interest 29

An American Triangle: Media, Violence, and Individual Responsibility 39
Blood on Their Lenses 27
Compliments of a Thief 47
Dishonesty: The Path to Public Confidence 12
Enquiring Minds And Space Aliens 3
For Sale: Justice—O. J. Style 66
Long Island Lolita Meets the Journalistic Lowlife 23
Make Mine Media Rare 82
Mixing a Bitter Pill 100
Playing the University Grantsmanship Game 70
Politically Incorrect Weather 35
Promises, Promises 10
Stripping Off Their Royalties 49
The Intruder 30
The Write Stuff 53
Three Nights a Week 126

Government

A Columnist's Best Friend 16
A Dam(n) for the Home Folk 21
An American Triangle: Media, Violence, and Individual Responsibility 39
An Obscene Story 84
'And Now a Word From Your Local Sponsor' 140
Conventional Fund Spending 117
Dishonesty: The Path to Public Confidence 12
Don't Despair, GM; Help Is on the Way 98
Dysfunctional Thoughts 68
Enquiring Minds And Space Aliens 3
Kernels of Truth 150
Make Mine Media Rare 82
Mixing a Bitter Pill for America's Drug Companies 100
President Sneezes; Film at 11 19
Promises, Promises 10
Selling Out America by the Yard 102
Spinning Their Way to Success 8
The Case of the Missing Popularity 6
Voices of America 130

War in the Gulf:
Lessons of an Obsession 142

Health

Cramping Up for Health 153
Kernels of Truth 150
Mixing a Bitter Pill for America's
Drug Companies 100
Tanness, Anyone? 152
The Case for Sensationalism 80

International Relations

Make Mine Media Rare 82
Rerunning America 55
The Press Meets the Afo-a-Kom 77
War in the Gulf:
Lessons of an Obsession 142

Labor Issues

An Hour a Day 186
Downsizing Baseball's Problems 175
Laying Off Marshbaum 180
Replacing Baseball 177
Traded for Two Rookies, an Editorial Clerk,
and a Future Draft Choice 178
'We're Management:
We Don't Have to Tell You Anything' 188

Language

Linguistic Larceny 59
'Local Reporter Cops Lexicon Story' 57

Licensing

Make Mine Media Rare 82

Media Ownership

Traded for Two Rookies, an Editorial Clerk, and
a Future Draft Choice 178
'We're Management; We Don't Have to
Tell You Anything' 188

Media Research

Playing the University Grantsmanship Game 70

Military

A Medal for the Army 146
A Major Conflict of Interest 29
Chicken Advertising 148
War in the Gulf:
Lessons of an Obsession 142

Obscenity

An Obscene Story 84
Of Matrons and Movies 86

Politics

A Columnist's Best Friend 16
A Dam(n) for the Home Folk 21
Dishonesty: The Path to Public Confidence 12
Electing the News 14
Enquiring Minds and Space Aliens 3
Promises, Promises 10
Rerunning America 55
Spinning Their Way to Success 8
The Case of the Missing Popularity 6
The Joy of Sax 18
Voices of America 130

Privacy

Long Island Lolita Meets the
Journalistic Lowlife 23

Public Relations

A Medal for the Army 146
Chicken Advertising 148
Disneyland: Where Dreams Are Real 109
Promises, Promises 10
Spinning Their Way to Success 8
The Case of the Missing Popularity 6
War in the Gulf:
Lessons of an Obsession 142

Recreation

Cramping Up for Health 153
DATELINE: London Bridge, Arizona 105
Disneyland: Where Dreams Are Real 109
Tanness, Anyone? 152
The Day the Circus Came to Town 103

Reporting Practices

A Columnist's Best Friend 16
A Dam(n) for the Home Folk 21
A Major Conflict of Interest 29
All in Vein 64
All the Beautiful People 159
Blood on Their Lenses 27
Creating Historical Fiction 75
Don't Despair, GM; Help Is on the Way 98
Dysfunctional Thoughts 68
Editing the 12 Commandments 63
Electing the News 14
Enquiring Minds and Space Aliens 3
Escalating an American Paranoia 42
For Sale: Justice—O. J. Style 66
Linguistic Larceny 59
'Local Reporter Cops Lexicon Story' 57
Long Island Lolita Meets the Journalistic Lowlife 23
Make Mine Media Rare 82
Politically Incorrect Weather 35
President Sneezes; Film at 11 19
Rerunning America 55
The Case for Sensationalism 80
The Intruder 30
The Only Alternative for KFAD 25
The Press Meets the Afo-a-Kom 77
Traded for Two Rookies, an Editorial Clerk, and a Future Draft Choice 178
TV News 101: Intro to Makeup 33
War in the Gulf: Lessons in an Obsession 142
Withdrawing From an O.J. Overdose 192

Sexual Issues

An Obscene Story 84
Jeanetics and the Cool 'Wannabe' 165
Long Island Lolita Meets the Journalistic Lowlife 23
Of Matrons and Movies 86
Sex and the Single Beer Can 168
There She Is, Miss-representing America 162

Sports

Downsizing Baseball's Problems 175
Replacing Baseball 177
The Monday Night Football Game 119

Technology

America's Ding-a-Lings 90
Curbing the Paperless Explosion 96
The Beeper Cacophony 92
The Impersonal Society 94

Introduction

Hardly anyone admits reading the supermarket tabloids, but someone—other than movie star publicists who "leak" information to the tabloids to create controversy—must be reading them because the combined circulation for the six major weekly newspapers is more than 10 million. I admit I am the someone who reads the tabloids, although it has been quite awhile since I read all 10 million copies in one week.

Usually, I read *The National Enquirer* and, occasionally, one of the other tabs. I seldom read the *Weekly World News* because in the *USA Today* world of splashy color and flashy graphics, the *Weekly World News* black-and-white front page just doesn't measure up.

Nevertheless, it was a hot August afternoon when I went into the local air conditioned supermarket to cool down and, perhaps, find a few of the 30,000 advertised items that could translate into dinner for six, including two German Shepherds. (The pot-bellied pig came later.) Apparently I wasn't the only one that afternoon who figured out how to get free air conditioning. The checkout lines were longer than a politician's lies, so there was only one thing to do. I figured I'd be able to get through most of the 15 magazines and six newspapers in the "you-gotta-buy-this" point-of-purchase racks by the time it was my turn. I also figured that my kids would have graduated from college, moved out of the house (something I probably will never see) and had grandchildren by the time I finished checking out.

Thus, it was in the checkout line that I learned from the *Weekly World News* that a space alien had come to earth to advise presidential candidate Bill Clinton. The alien had already advised President Bush and Ross Perot early in the Summer, but had to wait until after the Democratic convention to find out which of the donkeys was going to run. Because it was cooler earlier in the Summer, I hadn't had a chance to read those previous issues.

Being the alert reporter I am, I was upset that a competitor had scooped me on what could have been the most important news of the week. Just a couple of weeks earlier, I had covered the first Clinton-Gore bus tour of America when it came into Pennsylvania, and no one mentioned anything about an alien. Obviously, the Secret Service had covered it up once again.

That evening, my nephew from Georgia called. He had just read the space alien article, and knew I would be interested. The evidence was overwhelming. There were now at least two people who recognized good journalism. It was time to act.

For a few years, off and on, usually when I had too much time and not enough sense, I thought about writing a weekly newspaper column. It would be a great catharsis of what I proudly knew to be a warped mind, fertilized now and then by my wife. With only 23,000 other columnists trying to pitch their own catharses, I figured there was room for another 700-800 words a week, especially since newspapers appeared to be desperate for features; how else could anyone explain why they publish gossip columns and capsulized summaries of soap operas?

Thus was born "Wanderings," a weekly column that probes a small particle of society. Sometimes it's biting satire; sometimes a wistful essay. Sometimes it looks into politics, other times the environment, health care, recreation, or whatever needs to be probed that week. About half of the columns have a media focus. Occasionally, the media is the central focus, sometimes a supporting player, often an extra. But the media are always there—lurking; annoying; but most importantly, informing, analyzing, and entertaining—just as in real life.

Enquiring Minds and Space Aliens is a compilation of many of those media-related columns. Most of the columns have been revised for book publication. After all, newspaper columns stay around a couple of days, while books remain on the shelves, unread, for decades.

The columns have a logical order; for those readers who can figure it out, you definitely need to find a hobby. For the others, just read the columns. A few now; a few later. No one will rat on you if you read them out of order or if you fall asleep while reading the one column that has the secrets of the universe.

To assist you, I have included three tables of contents, all of them guaranteed to be as much fun to read as legal notices. The first Contents section is merely a running list of what's in the book. But, for those more adventurous—and for those humorless university professors who believe everything must be categorized and have a scholarly focus—the other two tables of contents define the columns by medium and by social issue. Hopefully, that'll shut them up while the rest of us enjoy life as it is.

Nevertheless, no matter how you read the book, my wish is that you find yourself not only entertained, but mentally stimulated and ready to act against the stupidity and injustice in this world.

—WALTER M. BRASCH

Enquiring Minds AND Space Aliens

Enquiring Minds and Space Aliens

When I don't believe I'm getting all the news from the nation's 1,586 daily newspapers, 7,417 weeklies, 11,000 radio and TV stations, 12,000 magazines, or 120,000 new books published every year, I turn to the supermarket tabloids for the truth. After all, it was the 3.1 million circulation *Star* which first revealed the existence of Genifer Flowers, Bill Clinton's alleged pre-presidential playmate.

Since most of the tabloid reporters have journalism degrees, worked on major daily newspapers, and are now earning $60,000-$100,000 in their new assignments, I place great credibility in what is being reported in the six major tabloids, all of them published in Boca Raton or Lantana, Florida, and which now have a combined circulation of about 10 million.

From the tabloids, I can monitor where Elvis is this week, learn first about who is being seen with whom, and which TV series is planning to replace which megastar, and more than anyone ever needs to know about soap stars, none of this reported by our usually vigilant local press.

I also know everything there is to know about Elizabeth Taylor, the Kennedys, the British royal family, Big Foot, and why taking coffee bean baths can perk you up.

I have also learned about monkey-faced boys, dog-faced girls, human-faced pigs, an 8-year-old who gave birth to twins, a woman who gave birth to a litter of 12 children, a 28-year-old grandmother, a man who was pregnant, and a tribe in South America that found a cure for cancer.

From the 350,000-circulation *Sun*, in one week alone, I learned that a survivor of the Titanic spent 20 years on an iceberg, that there really is a flying elephant with jumbo ears who lives in Zaire, that a woman is turning into Marilyn Monroe, that scientists in

Jerusalem found Goliath's mummified head, and that miracles occur near a Florida tree that has the face of Christ. The establishment media also don't report much about house hauntings, psychic revelations, reincarnations, and extraterrestrials. However, all conscientiously reported by the tabloids, and all for a buck or so a week.

In the current issue of the 722,000-circulation *Weekly World News*, I learned that condoms cause breast cancer, that a 7,000-year-old gargantuan shark patrols Lake Superior, that Hitler was really a woman who survived World War II and died in 1992 in Buenos Aires at the age of 103, and that a spaceship (with 14 perfectly preserved extraterrestrial corpses) was found in the Gobi Desert.

More importantly, I learned that a friendly space alien, not too unlike E. T., declared his (her? its?) support for Bill Clinton. A photo on page 1 showed the smooth-skinned, large-headed, long-fingered, unclothed alien shaking hands with the Democratic Presidential nominee after a 40-minute super-secret visit in Madison Square Garden during the Democratic National Convention.

It wasn't the first scoop for the *News*. In May, the newspaper had reported that the alien visited George Bush at Camp David; in July, it reported the alien stopped by Dallas for a chat with Ross Perot who, apparently taking the alien's advice, soon dropped out of the race. Pictures also accompanied these articles, thus proving the alien's existence.

We learned that the alien—who came from the most successful planet in the universe—gave Gov. Clinton advice on health and environmental issues as well as how to turn the economy around. The alien's mission—other than to evaluate and recommend a candidate for the confused American masses, most of whom would vote for the alien over any of the Presidential candidates—was to seek "trade concessions that would benefit his home planet," according to reliable sources who talked with the *Weekly World News*.

However, a few of my more cynical journalist colleagues, obviously jealous that they were scooped on the biggest news story of the decade, called the story a hoax. To get to the truth, I made a few phone calls. A member of the White House staff said she believed that the President made several light-hearted comments about the visit of the alien, but referred me to another office for confirmation. An official spokesman for President Bush at first indicated he didn't know what I was talking about when I asked about the space alien. After informing him of this late-breaking news, he said he didn't think the President made any comments about "that alleged meeting." He then informed me that "as far as we're concerned," there was no meeting, thus confirming my belief that if the White House says it didn't occur, it probably did occur.

On to Bill Clinton's team. The Governor had previously acknowledged that he discussed certain issues with an alien, and was pleased to receive the alien's support, an indicator that the Clinton-Gore team was broadening its base. At least that's what a few media reported.

But, being the hard-hitting investigative journalist that I am, I had to get actual confirmation, if not from the Governor, certainly from an official spokesperson. Did you ever try to find an official spokesperson when you need one? After three days of phone calls, all I had was a lot of conversations with a pack of confused but obviously arrogant campaign officials who couldn't or wouldn't confirm or deny anything. Obviously, the Clinton team was more impressed with themselves and in not revealing the truth.

To clear up the confusion, I finally contacted Eddie Clontz, editor of the *Weekly World News*. Eddie's a pleasant fellow and an excellent journalist who worked for the AP, then for several years as a reporter and news editor of the *St. Petersburg Times*.

"We had been working the story for a year," says Eddie who revealed that the newspaper received the tip from "some of our people in the military." He says that credible sources "often don't call regular newspapers because the dailies take it as a joke or will treat it as such," thus confirming my suspicions that daily newspaper reporters are more concerned with trivial things like crime and city council meetings than they are with news of interplanetary consequence. The photos, Eddie says, were submitted by one of the newspaper's sources. A true journalist, he wouldn't reveal a confidence.

He did confirm that the newspaper plans to follow the alien's travels through the country, but probably won't be tracking either George Bush or Bill Clinton.

"We don't get into political coverage unless it has to do with a space alien," the newspaper's editor says.

Now, for the big question. Does Eddie Clontz, editor of a newspaper with larger circulation than all but the top five American dailies, believe in the alien? "I really don't think so," he says, noting that although "the photographs look real to me, as a skeptic I'd say it's not true." Actually, he also calls the existence of the alien "preposterous."

In every political campaign, there is always something to break the tension, something to lighten up a campaign that tires out can-

didates, staff, reporters, and voters. This year, it's the alien's visit with the candidates. Next campaign, maybe it will be coverage of the alien's race for the Presidency.
AUGUST 1992

The Case of the Missing Popularity

It was late autumn when George Bush suddenly awakened after a deep sleep and panicked. His Popularity was missing. He knew he had it shortly after the Gulf War. He even remembered having it in the early part of the Summer.

So, one fine morning, he set out in search of his Popularity. The first stop was at the Lost and Found department of the Kennebunkport Emporium and Political Meeting Place which was having a Going Out Of Business sale.

"O.K., what was it?" asked the minimum-wage clerk. "A bill-fold? Your watch? Your kid? What did he look like?"

"Popularity," the President said. "Lost my Popularity. Miss it badly."

"What did you do to find it?"

"Looked around. Not there. Had it a long time. Popularity thing is gone. Bill Clinton stole it."

"Now, sir," said the clerk, "if you think it was stolen, and you think you know who stole it, you should report it to the police."

"I tried to, but the precinct's in the inner city. The Secret Service said it was too dangerous for them to go there."

"Look, fella, the store's shutting down in six hours. My rent's due, the finance company just took my car, my husband's been out of the army six months and still can't find a job, my kid's sick, and we don't have any health benefits."

"Wow! What great family values! You didn't get an abortion, and you're married! That's good!" The clerk looked at him and sighed, but still took a Missing Popularity Report, promised to let him know if anything turned up in the next six hours, then suggested he contact the press.

"Going to go to the typewriter place," the President told his Aides. "Going to get a typewriter and write a press release." At the Smith-Corona factory, he was confronted by locked gates and a sign that said, "Moved to Mexico."

"All *right*!" said the President. "I'll go to foreign countries, stop Communism, and find Popularity." Ten days and 35 countries later, the only thing the President learned was that his Popularity was ambushed in the Iran-Contra scandal, shot at by American-made artillery in Iraq, and wounded by the languishing American economy. And, it was still missing.

"I'll put an ad in the paper!" said the President enthusiastically. But, here he faced an even greater problem. Faced by mounting losses from a lack of advertising, the newspaper had put a "30" to its publishing. Nevertheless, there was hope. The former newspaper now housed a printing plant. The President figured he'd print up thousands of Missing Popularity flyers and post them all over the country. He figured wrong.

"Will that be cash?" asked the owner.

"Don't have that cash-thing right now," said the President. "Running a slight deficit. Had to buy socks last December to stimulate the economy."

"No cash; no flyers. Maybe you know someone at a savings & loan who can help you."

The President's next thought was to ask the army of homeless and the unemployed to help him.

"It's too bad there's only 600,000 homeless, and a few million or so unemployed," said the President. "If there were more, we could find Popularity even faster."

"If we wait a while," suggested one of his Aides, "we *will* have more who can help. Unfortunately, even if we get more, we can't afford to pay them."

"Pay them? I thought they'd do it out of love for their country. For patriotism. For God and family values. For the chance to be one of my thousand points of light."

"Only volunteers who think they'll get a plum after the election do anything for free," said the Aide.

"I'm the President! We'll draft the unemployed and homeless. Put everyone on TWA, fly into Flint, Michigan, commandeer cars, launch an all-out assault, finish up with a Hail Mary maneuver and find the Popularity." Of course, TWA was in bankruptcy, nothing struck fire in Flint, and it had been years since a draft was felt in the country.

"There's got to be some regulation that requires people to search for Popularity," said the President.

"Sir," said an Aide, "if you remember, in the past 12 years we deregulated everything. Banks. Radio. Airlines. Civil Rights and worker benefits. Even the environment."

"But I didn't cause the problems!" said the President frantically. "I was out of the loop. It all trickled down upon me."

"That may be true, but it did explode under your custodianship of the country."

"Big Business! They always supported me. They'll be my army!"

"Frankly, sir, they're quite selfish and support whoever can do the most for them. With your Popularity missing, they've begun cuddling up to the enemy."

That night, the President turned off the lights in the White House, vowing "Tomorrow is another day."

At that moment, over the White House, unseen even by the most sophisticated radar, was the President's Popularity, clutched firmly in the beak of a spotted owl flying south.

OCTOBER 1992

Spinning Their Way to Success

With his popularity lower than plankton on the food chain, the President had to do something dramatic if he had any hope of a second term. Something came in the form of a top secret meeting of the White House Office of Presidential Spin (WHOOPS).

In the Silverado Room in the West Wing one Monday morning, WHOOPS planned to spin the President into reelection. With only a few weeks left before they were unemployed or, worse yet, had to return to jobs in advertising, the spin doctors were desperately searching for solutions. Since the President's popularity was at its highest during the Persian Gulf War, it wasn't unusual for the first solutions to center around the War. The fact that there was little discussion about Sadaam still being in power, that the Kurds were being massacred, or that there was still no democracy in Iraq, Kuwait, or Saudi Arabia was already a tribute to the effectiveness of the President's image makers. But it wasn't enough.

"We need something new. Something dramatic," said Dr. Clement P. Fielgut, the chief doctor of WHOOPS.

"We nuke the Soviets," said Sally Sweetwater. "No one likes them Commies anyhow."

"There's no more Communists," said Dr. Fielgut sadly. "Didn't you read the papers?" As soon as he said it, Dr. Fielgut knew he made a mistake since Sweetwater had been a TV anchor before moving to Washington.

"What about China?" countered another member of the staff. "We declare we're saving the world supply of Wedgewood dishes?"

"Wars are out," said Dr. Fielgut. "Besides, in the past 12 years, we invaded every land mass on earth that has oil or Communists."

"We launch a battle cruiser to Mars!" declared another doctor.

Dr. Fielgut just looked at his staff, knowing they were Presidential appointments and the result of inbreeding among financial contributors.

"We could focus on something the President did the past four years that benefitted mankind, then exploit *that*," suggested one tactician. After the laughter died, the faceless manipulators tossed around all kinds of ideas. A redirection of national priorities. A larger information program. A national health care plan. An increase in the number of social workers with a concurrent decrease in personnel in the executive management branch of the federal government. Naturally, all the ideas were seen as unworkable and thrown out.

"What is it that we do here?" Dr. Fielgut asked rhetorically.

"Modify the truth, of course," came the unified reply. A light flickered in the mind of Lance Redux, a recent honors graduate in public relations, and the newest member of the WHOOPS staff.

"If we can convince the people that the President was the one responsible for curing AIDS," said Redux, "we'd be able to make people forget about Irangate, poverty, the health care crisis, crime, drugs, and even Dan Quayle. The President would be a shoo-in not just for reelection but for the Nobel Prize as well."

"It could work," said Dr. Fielgut, "but first we'd have to make up an AIDS policy."

"Remember back seats of cars?" Lance Redux asked, patting down his lacquered hair. "Remember how wide and soft they were?" They remembered. "What happened in the back seats?" he asked.

"I don't know how she got there!" one doctor shrieked.

"It wasn't mine!" squawked another.

"I wasn't even in the state at the time," burbled a third.

However, the junior spin doctor advised the doctors to remember the Recession, and that it could be used for the President's benefit.

"What Recession?!" thundered Rockhead, one of the doctors. "There never was a Recession! There is no Recession! There never will be a Recession!"

"That's right," said Dr. Fielgut, explaining a version of reality to the newest staff member. "We had a downward modification of an upwardly spiral economy. There may have been some temporary lapses of continuing income maintenance. But we *didn't* have the "R" word."

"We do now," said Lance Redux. "We need to identify the past four years as a Recession. We need to exploit it. We need to make everyone now believe that the President was responsible for the economy."

"Treason!" shouted one member.

"Sedition!" proclaimed another.

"Heresy!" the others ordained.

But Lance Redux pleaded for just a minute more. The others granted him his one last request.

"What happens in a Recession?" he asked.

"People can't buy anything. Any fool knows that!" came a hostile reaction.

"And when people can't buy anything," said the newest spinner, "they don't buy cars. Even if they had downsized economy cars, they still couldn't afford the gas. And when they don't have gas and don't have cars, they can't have sex. And when they can't have sex, they can't get AIDS."

On one fine Summer morning, 10 spin doctors (and one potential cabinet secretary) walked into the President's office to announce that because of the President's leadership, the only screwing that occurred in the country the past four years had nothing to do with sex. They had finally found something the President could take responsibility for.

AUGUST 1992

Promises, Promises

With less than a month before the election, Marshbaum was campaigning furiously.

"A chicken in every pot! Free health care for everyone! There's light at the end of the tunnel!"

"Marshbaum!" I commanded, "you can't make those kinds of promises."

"You're right. I don't want to offend the health care industry. There's a lot of campaign money there. I'll just make up something else."

"You just can't make up campaign promises."

"Sure I can. It's easy. How about 'Vote for Marshbaum and win a date with Bette Midler?'"

"You don't even *know* Bette Midler."

"I like her movies," he said casually.

"It has nothing to do with her movies," I said.

"Think someone doesn't like her singing? I sure don't want to offend anyone. I could make it a date with Natalie Cole. How about Pat Boone for the women?"

"Marshbaum!" I screamed. "Get reasonable!"

He thought a moment. "You're right. Nat and Pat are probably supporting someone else. How about 'Elect Marshbaum and you'll never pay taxes again!' "

"That's ridiculous," I said. "No one will believe you."

"Doesn't matter if they do or don't as long as they vote for me."

"But you'd be lying to the people," I said.

"It worked for Sen. Packwood," said Marshbaum smugly. Not long after Bob Packwood of Oregon was elected to his fifth six-year term, he admitted he had lied during the campaign when he denied he had made numerous improper sexual advances to members of his staff during the previous 20 years. A group of voters had petitioned the Senate to overturn the election on the basis that Packwood had lied to the people. The Rules Committee thought about issues of the greater public good, then they remembered their own campaigns, and unanimously declared that lying to the people during a campaign wasn't strong enough grounds to overturn an election.

"There's got to be some law that prevents politicians from lying," I said.

"Even for being a journalist, you're rather dense," said Marshbaum. "Not only is history written by the victors, but Hitler himself said that the victor will never be asked if he told the truth."

"I doubt that quoting Hitler is going to get you very far," I said.

"It's not *who* said it," said Marshbaum smugly, "but the truth of *what* he said. Besides, the FCC says it's OK to lie."

"The Federal Communications Commission gives its approval?" I asked skeptically.

"Section 315. The FCC says that radio and TV stations can't refuse to run political ads even if the station management knows the ads are outright lies."

"Most people don't believe most of what they see on TV anyhow," I sniffed.

"Try the Supreme Court," said Marshbaum.

"The Supremes said lying to the people is acceptable behavior?" I scoffed.

"OK, not the *U. S.* Supreme Court, but *A* Supreme Court."

"Which one? In Baghdad?"

"Albany. The New York Supreme Court."

"Marshbaum, not even New York's court could be that incompetent."

"Got it right here," he said, taking a wadded paper from his pocket. Case of *O'Reilly v. Mitchell.* Guy named O'Reilly sued a politician named Mitchell in 1912 and charged him with making promises that weren't kept."

"A promise is a verbal contract," I said. "I'm sure you read it wrong. The Court undoubtedly *upheld* O'Reilly's claims."

"Wrong, Newsprint Breath," said Marshbaum smugly. "The Court said that politicians lie all the time, that promises in a campaign are just that. Promises. Verdict for the politician. Case closed."

"But that occurred before World War I," I said.

"It's precedent," Marshbaum said. "It's on the books. How about 'Vote for Marshbaum and he'll wash all your dirty laundry?' "

"Marshbaum," I said disgustedly, "you can do whatever you want, but just remember that some politicians actually tell the truth."

"Name one who did and got elected!" he demanded.

"Honest Abe," I replied.

(For the legal scholars, the case of O'Reilly v. Mitchell *is cited as 85MISC176, 148NYS, 88 SUP, 1914.)*

OCTOBER 1993

Dishonesty: The Path to Public Confidence

Congressman Blatant Lee Pure paced the floor of his office, visibly upset, and worried about being unemployed in a few days. As the only unindicted congressman, Pure's ability to lead his district was being questioned.

For awhile, the people were so upset with the scandals of Congressional politics they planned to vote everyone out of office. But, they soon realized that although their elected representatives were crooks and had little concern for the people, at least they knew how to toss them a few morsels from the pork barrel.

In Pure's home district, the people were defecting to his opponent, a man more crooked than the stream that meandered aimlessly through the center of the district. As one of Congressman Pure's former supporters accurately noted, although everyone knew Pure's opponent was a thief, "at least he'll get us things, no matter what they cost!" Even worse for Congressman Pure, the media ignored him, reasoning that any legislator who hadn't done anything to be indicted couldn't be worth writing about.

Back in his Washington office, Congressman Pure turned to his one remaining aide—the others had long ago quit in fear of being implicated in honesty—and announced he could learn to become dishonest. "I suppose I could run a traffic light," he suggested.

"That won't help," the Aide replied. "Everyone runs red lights in Washington. It's when you stop for them that the police get suspicious."

Congressman Pure tried again, knowing he had to do something not only illegal, but that would make headlines as well. "I'll steal a lollipop from one of the congressional pages!" he suddenly shouted.

"Congressman," said the Aide patiently, "I don't believe you have the idea. It's not good enough to run a red light or take candy from a kid. Everyone does those things. You have to do something to make the media stand up and take notice. When that happens, the voters will again turn to you for the inspired leadership they've come to expect from members of Congress. In short, you have to be bought!"

"You mean like Congressman Fishbender?"

"Sure. That old goat traded so many votes that when the armed forces appropriation bill came up, it had a new Navy base in it for his home district. In fact, it's the best-looking federal installation in all of Iowa. Now, *those* voters are grateful people. Fishbender's been re-elected every election since."

"But my vote is sacred. I even make lobbyists see me in the office and not at a hundred buck a meal dinner. Besides, no one needs my vote."

"Think positively," said the Aide. "Maybe the lobbyists for each side could buy only half the congressmen. That's 217 each. That could leave you—the 435th one—as the swing vote. By the time you cast your vote, you'll have a car, a vacation cabin on 10 acres of fertile grassland, and possibly even a new highway for your district."

Congressman Pure thought about the alternatives, then decided that the only thing to do was to concede the election, unable to be a part of the noble traditions of Congress. But fate wouldn't allow it. The Congressman's honesty had left a blemish on the record of Congress, and it was up to fate to correct that blemish. Into the Congressman's office walked a newscarrier for the *Washington Post*, asking for $9.20 for the month's subscription.

"I don't have the money this week," said the Congressman, explaining he donated what was left of his salary to needy children, a program for the retarded, and then gave the rest away to the Red Cross for disaster relief. "Can you come back next week?" he asked.

The newscarrier looked at the Congressman, informed him

that the money was due, and that if the Congressman didn't pay, not only would the subscription be canceled, but the newscarrier would lose out on a free trip to Disney World for selling the most newspapers.

"But, I don't have the money," the congressman again pleaded.

Faced with losing the free trip and a chance to sell a subscription to Mickey Mouse, the newscarrier did some fast thinking and paid for the subscription out of his own pocket. Naturally, such a charitable act was bound to make the news. Within a week, Congressman Pure's hometown folks rallied to support him, confident that they finally had an important Congressman. And all of them had copies of *The New York Times* which had dutifully reported, in bold multi-column headlines–"*Washington Post* Buys Congressman; Sources Say Congressman Pure Tool of D.C. Paper."

And, after all, if one is to be bought, isn't it better to be bought by the media who create the images than by the lobbyists who create the problems?

SEPTEMBER 1992

Electing the News

Days before the election, Picapole was fighting hard to retain his seat as editor of the *Daily Noise*. Challenging him was Leadshot, a law school graduate who had worked his way up from janitor to president of his father's gun manufacturing company.

The latest polls showed that in the two-county area the *Noise* served, Picapole and Leadshot were in a virtual dead heat, with an additional 10 percent of the vote, mostly reporters, aligned against having any editor.

First elected in 1984 as a fresh candidate against trickle down journalism, Picapole had been comfortably re-elected every two years since. But now, with the readers demanding change, Leadshot had charged from 15 points behind, and was on the verge of winning his first election since the 6th grade when he became assistant hall monitor.

"I'll vote for anyone running against Picapole," said a determined Marvin Blunderbuss of Porkbelly, Pennsylvania, who admitted he never read the newspaper, but defiantly insisted that anyone who spends more than four years as editor is part of the establishment and must be replaced.

In a slick TV ad campaign, financed primarily by the American Medical Gun Association, Leadshot charged Picapole with being soft on reporters, and blasted him for allowing two reporters to leave the *Noise* before their contracts expired. "These scumbags," said Leadshot, "are now writing award-winning stories for the *Morning Blab*, and are killing us with their constant scoops."

Picapole acknowledged it may have been a mistake to allow the reporters to break their contracts, but claimed that in the 10 years he was in office, the *Noise* efficiently picked up its reporters on the streets right after committing a journalism degree, and sentenced them to work slave-like conditions until the end of their terms.

Taking the offensive, Picapole charged that Leadshot himself was soft on reporters. Waving a sheaf of statistics, Picapole pointed out that for all his histrionic blustering, Leadshot allowed his own staff to be pummeled by reporters 83 percent of the time, and that if he were to be elected there would be "every evidence that he will continue to preach toughness yet be soft on news."

Leadshot countered that at least if he were elected editor the newspaper would run more good news stories. Picapole retorted that the media can't make up hard news when there isn't any.

Picapole then charged that Leadshot not only was a poor businessman, but that the readers shouldn't be swayed by anyone who never worked on a newspaper, never even took a journalism course, and wouldn't know the difference between a hole in news copy and a hole in his assets. In response, Leadshot scored points when he retorted that the biggest problem with journalism today is that it's being run by journalists.

Not letting up on his attack, Leadshot produced 8-by-10 glossies of a press party in which Picapole was seen standing near the editor of *The New York Times*. Picapole responded that although both he and the *Times* editor were members of the same party, they barely knew each other, and that he would never lower himself to ask for an endorsement.

"The economy is in a dumpster," Leadshot charged in a major speech in Hogswallow, Pennsylvania, "and Picapole is responsible for it with all his spending on reporter salaries and expenses." Picapole acknowledged that in the past 10 years he was responsible for increasing reporter salaries by 3 percent, bringing them just past the poverty line, but said he had no choice in the matter of buying wrist rests and non-glare computer screens since the newspaper lost a suit when 11 of his nearly-blind reporters with carpal tunnel syndrome won a class action suit in federal court.

In one of his more popular yet controversial planks, Leadshot also proposed that every reporter be required to carry a gun, a proposal that split the newsroom's loyalty. "You never know when your life will be threatened at a school board meeting," said a long-time reporter who asked that for matters of security he not be identified. However, another reporter who also asked for anonymity since she feared retribution from the Leadshot faction, said she would oppose any reporter being required to carry a gun. "It'll just escalate problems," she said, then summarized a report that revealed that reporters who carry guns are five times as likely to be shot by their editors as are unarmed reporters.

Nevertheless, throughout the election the readers have been puzzled by the lack of substantive debate about issues of news design over substance, artificial limits on story length, the candidate's response to the lack of resources and personnel to cover the news, as well as a general lack of health, labor, and environmental reporting. A recent poll shows that 94 percent of voters are making plans in two years to repaginate whoever wins.

OCTOBER 1994

A Columnist's Best Friend

I guess it can now be told. Those of us who write newspaper columns are going to miss Dan Quayle. Every time we got writer's block, every time there was a slow news day, there was always Dan Quayle to pull us out. Many of our shots were cheap, but at least they filled space and kept our editors happy. Many of our comments about the Vice-President even made it onto national newscasts and were repeated by the nation's comedians.

In the spirit of fair play, Dan Quayle wasn't the complete child-like inattentive carefully-manicured golf-swinging draft-evading academically-average airhead many thought he was. If we're completely fair about it—and everyone knows no newspaper columnist is ever completely fair—we'd have to admit that the vice-president grew significantly during his four years. He effectively helped lead the fight for a product liability bill. During the attempted coup in the Philippines he ably directed the National Security Council while the President was on his way to a summit conference in Malta. And, most importantly, the vice-president spoke out on issues, intelligently challenging his President to consider other options. It was something not even George Bush did as vice-president. Certainly, Mr. Quayle's courageous attacks on lawyers and on a decline of family values opened up national discussion on issues that needed to be brought before the people.

But, then, there was still the quintessential Dan Quayle who was what we satirists looked for to bail us out.

It all began in 1980, shortly after his election as a U.S. senator.

"I know one committee I don't want—Judiciary," said the graduate of the Indiana University law school. "They are going to be dealing with all those issues like abortion, busing, voting rights, and prayer. I'm not interested in those issues and I want to stay as far away from them as I can." It ended with unwed TV journalist Murphy Brown (she's real isn't she?) affectionately telling her newborn baby that, like all babies, he had a partially developed brain and could therefore become a vice-president. Inbetween was a lot more fun.

A few months after the 1989 inaugural, the vice-president tried to help the United Negro College Fund by repeating its slogan that "a mind is a terrible thing to waste." Of course, in Quaylese, it became, "what a waste it is to lose one's mind. Or not to have a mind." He also tried to praise America's teachers by revealing, "Quite frankly, teachers are the only profession that teach our children. It is a unique profession." He called American Samoans "happy campers," and during the declining days of the Soviet Union declared that "hatred of God" was the reason for that nation's domestic problems.

He told the delegates at an NAACP convention that he was "committed, just as George Bush is committed, to fostering good and better relations with the civil rights community. The President and I are on your side." Kinda makes us all want to rush right out and buy insurance.

But, he was also a golfer. Over the years, he probably played at a lot of country clubs that excluded Blacks, Hispanics, Latinos, and Jews. It caught up with him in Pebble Beach, California, where the Professional Golfers Association had just forbidden Cypress Point County Club from hosting a sanctioned tournament because the club didn't have any minority members. Virtually every newspaper in the country ran the story, including sidebars and follow-ups. Then Dan Quayle showed up and played a round with the secretary of the Air Force and a U.S. diplomat who were club members. Mr. Quayle did cancel the second round, saying, "playing there might look bad," that he hadn't read any of the sixteen quintillion articles, that he didn't think the club deliberately discriminated, and reaffirmed his position that he believed in civil rights.

Of course, civil rights didn't apply to women since he said he saw no reason to cancel his membership in the all-male Burning Tree Country Club near the nation's capitol. "Maybe they'll change," he said. "I think it would be a good idea for them to take women into the club. I don't have any problem playing there in the meanwhile."

About the nation's somewhat controversial invasion of

Panama, the vice-president stood behind his President. "I'm convinced that the President's correct and courageous decision to do what he had to do in Panama will not be of long term consequences in a negative sense," said the vice-president. Shortly before the Gulf War began, he fearlessly declared, "We are ready for any unforeseen event that may or may not occur."

Finally, during the re-election campaign, in which health care reform was part of the national agenda, Dan Quayle tried his best to explain why Hawaii's health care program, which many political leaders were praising, may not be relevant to the rest of the nation: "Hawaii is a unique state. It is a small state. It is a state that is by itself. It is a—It is different than the other 49 states. Well, all states are different, but it's got a particularly unique situation."

Yes, we're going to miss Dan Quayle. But, we need not feel too sorry for him. After all, when was the last time you heard of an unemployed relative of a newspaper owner?

NOVEMBER 1992

The Joy of Sax

There aren't many people more thrilled than I am about President Clinton's week-long inaugural festivities.

After 12 years of abstinence during the Reagan-Bush era, but now inspired by the new administration, I am planning to go to my local music store, buy a few reeds and cork grease, and resume saxual activity. I'm frightened by the possibilities, but I know it's the right thing to do.

At one time, saxophones had defined the sound of the "Big Band" era, and then were the driving beat in the beginnings of rock. But, during the Reagan-Bush years of deregulation, saxophones could no longer compete against instruments that were picked, plucked, and strummed, and became just another instrument whose sole purpose seemed to be to provide a musical bridge in high school marching bands between the ranks of trumpets and clarinets.

Unable to get steady work as a rock saxophonist, I packed away my silver alto sax and golden tenor sax into their separate but equal cases, in keeping with the re-emerging American philosophy.

Into the "me era," a generation of Americans shoved music and the other creative arts out the back door while escorting greed in the front door. The Administration's policies of the unregulated free market were destroying sax. Then, when the administration declared family values were more important than problems created by Irangate and Iraqgate, the savings and loan debacle, unemployment, American corporations moving to other continents to exploit a whole new class of workers, a recession, wars in the Balkan, starvation in Somalia, and even a war in the Persian Gulf desert to defend the sanctity of oil in countries more sexist and racist than anything America ever envisioned, then turned its back on the suffering of the Kurds who didn't control oil, we knew sax was no longer important. But, secretly, I wondered how families could survive and regenerate themselves without sax in their lives. Although the nation had committed numerous indecencies against itself, what the administration did those 12 years was nothing short of saxual harassment.

While the Republicans were pushing family values, the people of Arkansas were quietly applauding (or at least tolerating) the rebellious Gov. Clinton's saxual escapades. But, most of us didn't become aware of Clinton's obsession with sax until he grabbed his instrument and a pair of shades and blew out "Heart Break Hotel" on the *Arsenio Show*. That's when I knew that I had to vote for Bill Clinton. Anyone that brave, who was willing to stand up for his beliefs and go onto national television to endorse the creative arts was going to get my support.

So now I'm watching the inaugural festivities, listening to my fellow Americans honestly tell me about their saxual orientations, and hoping to resume where I left off a dozen years earlier. Only this time, I plan to buy a mute and practice safe sax.

JANUARY 1993

President Sneezes; Film, at 11

Whenever the President sneezes, 75 reporters file their stories. Because it's the President, and because it's easier to file a Presidential Sneeze story than to root through the causes of the health care crisis, the reporters fall all over themselves trying to beat each other out to get the comprehensive coverage TV-News fans have come to expect.

Anchor: This is Clyde Barrow at the White House. The President sneezed about 2:45 this afternoon. We understand it lasted about three seconds. We now take you to the emergency room of the Bethesda Naval Hospital. Standing by is Lance Redux.

Lance Redux: The President has just arrived, and we'll be interviewing bystanders, orderlies, and maybe even a nurse or two. Security is extraordinarily tight, and only the 237 accredited reporters have been allowed into the ER at this point. Anticipating a Presidential sneeze, and to give more room to the news media, every area hospital for the past week has been sending patients without health insurance to the Fumigate Center in Arlington. Back to you, Clyde.

Clyde Barrow: With me is Gatekeeper Jones, special assistant to the president on sneezing. Ms. Jones, what does this sneeze mean to the American people?

Ms. Jones: First of all, let me say that the President has made great inroads into understanding and responding to the needs of the American people. This sneeze was not only the result of his tireless efforts on behalf of the people, but also a declaration that he and the First Lady plan to go to extraordinary lengths to deal with health care issues in this great nation of ours.

Clyde Barrow: Thank you for that great insight. Now, to Susie Sweetwater with Sen. Porkbelly Pineapple at the Capitol.

Susie Sweetwater: Sen. Pineapple, we just heard that the President's sneeze was in sympathy with the plight of Americans everywhere. Do you agree?

Sen. Pineapple: While all us Americans are concerned about the President's health, this particular sneeze was the result of a President who has disregarded the wishes of the people and the Congress, who has thumbed his nose at the health care industry, and is foolishly going where no president has gone before—at least not for the past 12 years, that is.

Susie Sweetwater: For an opposing view, we turn to Rep. Horace Sludgepump.

Rep. Sludgepump: While I don't wish to disagree with my esteemed and most distinguished colleague from the other side of the capitol, I should point out that if it weren't for him and the other cretins from the opposition party who filibustered the death of so many of our great and glorious programs which were designed by our party to help the working class, we'd have a chicken in every pot in this glorious country.

Clyde Barrow (interrupting): Excuse me, Susie, but the President's personal physician is about to make an announcement. We now return to Bethesda Naval Hospital.

Dr. Alfred Chiu: After running a series of tests on the President's bodily fluids, examining X-rays of his nasal passages and respiratory system, and checking the results of the MRI-scan, we now believe the cause of the sneeze was a pollutant in the air.

As you know, the President has many allergies. We have not yet identified that particular pollutant.

Reporter 1: Harry Hotlips, ABC-Action News. Doctor, can you identify that pollutant?

Dr. Chiu: As I mentioned, we haven't yet identified that pollutant.

Reporter 2: Judy Jumpstart, CBS-TV. Just how serious is this pollutant?

Dr. Chiu: We can't determine how serious the pollutant is until we can identify it.

Reporter 3: Darla Dazzling, NBC-TV. Doctor, just exactly what kind of pollutant could that have been? And does it have long-term effects?

Dr. Chiu: I don't know, but we will try to find out.

Reporter 4: Sid Serious, CNN. Doctor, could you indicate what you believe would be the world consequences of this particular sneeze.

Dr. Chiu: I can't at this time, but I will ask the Secretary of State to respond as soon as he completes his phone calls to the other world leaders.

Reporter 5: Polly Prattle, *New York Post.* Was the President sleeping with anyone other than his wife when he was polluted?

Clyde Barrow (again interrupting): I regret that we must temporarily interrupt our in-depth team coverage at this point. Right after these important messages from our sponsors, we'll return you to "General Hospital," already in progress.

JUNE 1993

A Dam(n) for the Home Folks

With the arrival of the new TV season, it's only a matter of time until one of the TV networks buys out C-SPAN for the rights to prime-time coverage of "The Congressional Follies."

"Quiet on the set! House of Representatives, Monday session. Take 1."

Speaker: I'm Troy Calhoun, speaker of the House and your host for Congressional Follies. But first, a message from our sponsors.

Sponsor: Are you tired of people taking advantage of you? Upset that your views aren't heard? The Ace Lobby Service has direct access to some of the people who can improve your life. Whatever your needs, contact Ace Lobby Service. As always, prices are negotiable.

Speaker: Welcome back. With us today is Rep. Howard Sludgepump of Oklahoma with a bill he'd like to tell us about.

Sludgepump: Thanks, Troy. And it's certainly good to be here with all you wonderful people. As I tell my folks back home in Hushaby Holler, Oklahoma—some of the finest people in the world—this is a real honor for me to present their wishes to the Congress of the United States. In fact, just the other day—

Speaker: Forgive me for interrupting, Sludge, but the floor manager just gave me the speed-up sign. Please don't keep us in suspense. What's in your bill?

Sludgepump: Dams. A billion-dollar dam right in the middle of Hushaby Holler, Oklahoma.

Speaker: Well, let's hear it for Rep. Sludgepump's billion-dollar dam. *(A chorus of cheers).* Now, ladies and gentlemen of the House, are there any questions you would like to ask Rep. Sludgepump? . . . Mr. Popoff.

Popoff: Thank you, Mr. Speaker *("How's my make-up. I hope they get my left front close-up. I look better that way.")* Mr. Sludgepump, according to the records of the Weather Bureau, Hushaby Holler hasn't had rain for 20 years. There are no rivers or lakes. And your district is the wealthiest in the nation because of all the oil you have. Why is that dam necessary?

Sludgepump: That's a very good question. And the folks in your home district of Wattabago, Iowa, can be assured that they've really got a good congressman looking out for their financial interests, especially since you convinced the Congress to open a federally-funded Museum of Corn in your district. Yes, sir, it's hard to find congressmen as concerned about the taxpayers' money as you are. So, all you folks in Wattabago, Iowa, made sure you vote the Popoff ticket in November. Are there any other questions I can answer?

Rep. Hotchkiss: Congressman, isn't a billion dollars a bit much for a dam?

Sludgepump: Not really, Mr. Congressman. After all, you have been doing a fantastic job for the large corporations in your district to help them find cheaper labor in Mexico, and know the importance of full employment. You, my most esteemed colleague, should be the first to realize that the billion dollars will be used primarily to feed underprivileged construction workers, PR people, and thousands of administrators who would be out on streets penniless and starving if we didn't pass this bill. Certainly, a man as courageous as you are in helping the people of your district who will be given preference for work on this dam can see that.

Speaker: Well put, Mr. Congressman. I'm sure Bushneck folks know that Congressman Hotchkiss really cares about them. But, right now, we must break for a public service message from the New York Stock Exchange.

Sponsor: A lot of people say that big corporations are capitalistic conglomerates, bent on making profits at the expense of the people of this country. We don't have any great conspiracies or plans to take over the country. We want to do only what's best for the country. And we want you to know it.

Speaker: Thanks, Dad. We're ready to vote. All those in favor of the dam bill, signify by removing your sunglasses. Those opposed, you may leave your sunglasses on, and hit the miniature gong on your desk . . . Well, it appears we'll soon have a billion-dollar dam in Hushaby Holler. This is Troy Calhoun wishing you courage for another tomorrow.

OCTOBER 1993

Long Island Lolita Meets the Journalistic Lowlife

It was 6 a.m., Monday, when the phone rang, so I knew it was Marshbaum eager to involve me in his latest scam. I wasn't disappointed.

"Got any extra money?" he whispered.

"Not since I became a humor columnist," I replied.

"Too bad," said Marshbaum. "I'll just have to take it elsewhere."

"Take what elsewhere?" I asked sleepily.

"Revelations about my life with Amy Fisher."

"*You* had a life with the Long Island Lolita?"

"Actually, I once bought some fabric at the sewing store her parents own, but I figure what they told me about their daughter is worth a few thousand on the journalistic market."

"What'd they tell you?" I asked.

"Not so fast. It'll cost you. You been looking at the tabloids lately? They pay real good for my kind of news."

"Look, Marshbaum," I said, "the papers I write my column for are respectable and ethical. Usually. They won't pay for revelations."

"Everyone else is!" he said outraged. "Three hundred thousand to the woman who got shot and her husband who either was or wasn't Amy's pizza-eating pimp, Fifty thou to some boyfriend. A bundle to a trigger-man. Quarter million to some hack to write a paperback about all this. Thousands to just about anyone who ever lived in the same county she does. Millions to produce the Amy Fisher Film Festival on TV. Add in a half-dozen tabloid TV shows, a few talk shows, and all the local news, and you have a billion dollar Amy Industry."

"But, that's schlock entertainment," I said. "Newspapers are more into news."

"Yeah," he said sarcastically, "like the New York papers that have photographers staking out everyone's houses? Or, the battalion of reporters who invaded her upper class hometown to talk to all the neighbors? What about the newspaper reporters who are being paid by the entertainment industry to be 'consultants'?"

"I have no desire to write about sex-crazed teenagers on the prowl for body-shop repairmen," I said. "Besides, every hour they take to report on this teenage prostitute with a Fatal Attraction complex is one hour less they can investigate corruption. Every inch they use in every paper is one inch less than can be devoted to stories about the economy and health care crises."

"What about a scoop on body-shop repairmen who lure teenage girls into lives of crime?" he asked. "I can sell that one a little cheaper."

"He may have been a sleaze," I said, "but there's no evidence he suggested that his alleged lover kill his wife."

"Precisely," said Marshbaum. "There's no evidence of any of this, but TV has already devoted more air time to it than they did to the Gulf War. Besides, you're the only one who hasn't written about it. Aren't your editors concerned that you're not giving them the latest news? Sin sells papers. Makes money for publishers. Publishers reward sinners by increasing their salaries. Frankly, you're well behind at the moment."

"OK, Marshbaum, what choice piece of trivia did you learn that'll save my career and get me the Pulitzer Prize for Sensational Journalism?"

"Like I said, I'm no fool. This stuff's too hot to entrust to someone not willing to pay."

"I said I had ethics."

"Ethics don't pay the bill at Victoria's Secret," he said. "Besides, it's relatively simple. You get the scoop on why Amy's parents never thought it was unusual that their 16-year-old daughter not only had a beeper, but a nearly-new usually bent-up Dodge Daytona with an automatic pilot to guide it into the body shop. You put it into your column. Some low-life publisher or producer with a wad of bills calls you up, pays you even more. Everyone makes out."

"Everyone but the readers," I said.

"It's the readers who want this stuff," he reminded me. "They'll read your column before they'll read about poverty and the health care crisis."

"That may be true," I said, "but I'll pass on this one. My readers will just have to remain ignorant of some of the significant events that affect their lives."

"Well, don't come crying to me when "Hard Copy" scoops you, and your editors replace you with someone known as 'a highly reliable source.' " I said I'd have to risk it. "By the way," Marshbaum said shortly before hanging up, "I'm sure you'll get a column out of this somehow. I'll send you a bill in the morning."

JANUARY 1993

The Only Alternative For KFAD

The news rating of television station KFAD went out to lunch and didn't return. This, of course, caused great concern for the news director who didn't like the possibility of being exiled to a small 500 watt radio station in Hogshead, Iowa.

"What about story length?" the station manager snapped. "Haven't they been getting longer?"

"No, sir! We've never run a story longer than 90 seconds, and most of the stories are 15 or 20 seconds."

"Good. We're still using the local newspapers, right?"

"Oh, yes, sir! My reporters know that the only way to get good television news is to rewrite from the local papers. That way we devote our energies to stories of real importance."

"Like that beauty pageant last week," said the station manager enthusiastically. "Real hard-hitting news coverage."

"It just lent itself to our newsfilm coverage. I sure hated to miss the Senate investigation, though, but since they don't allow cameras in the chambers, there wasn't any sense having a reporter there."

"Well," said the station manager reflectively, "there's nothing wrong with our basic news coverage, so it has to be our on-air personalities. How long has it been since Susie Sweetwater changed her hairstyle?"

"About two months, but I think we should let it stay that way another month or so just to increase viewer identification."

"In another month or so," said the station manager tersely, "we may be fourth in a three-station market. How's Susie handling the slump?"

"Remarkably well. Just yesterday, she had her fingernails manicured, and bought a new dress. I think she looks a lot better than Laura Landfill over at that other station. The technicians certainly have noticed."

"But, hasn't she presented a few special problems to your staff?"

"It's true our newswriters are cramped writing scripts with no more than two syllable words in them, but someone saw Susie actually trying to read a newspaper last week."

"What about Heartthrob? You know the co-anchor is just as important."

"He's been putting on a little weight, so I sent him over to the spa. It'll boost his image immeasurably."

"How's our ethnic balance? You know how the FCC is. Can we get better ratings by adding another ethnic?"

The news director shook his head. "We already have two Chicanos, three Blacks, an Alaskan, an Indian, two Laplanders, and a Southern Baptist on the news team. I don't think we can add any more right now."

"Maybe we can laugh it up more on the set. Y'know, maybe expand the Happy News format?"

"We're saturated now, and I don't think the viewers are ready to tolerate Susie giggling through another air disaster."

"What can we do about the weather?"

"As you know, McDonald gave the weather once in a raincoat, and another time in a bikini. And remember the times she brought the chickens into the studio to apologize for laying an egg with the previous forecast? You can't do much more than that to get ratings!"

"Pack your overalls," said the station manager. "It looks like you're heading to corn country."

"Just give me time," pleaded the news director, "I'll come up with some innovative way to boost the ratings." There was a moment of silence, then the news director blurted out his idea. "A journalist!"

"A what!?"

"A journalist! That could be the radical new way to boost the ratings. We could hire a journalist to give the news!"

The station manager was furious. "Don't be ridiculous! A journalist on TV would be a disaster!"

"No, Boss, I mean it. No one else has a journalist on the air. It could be the novelty that sells the station to the advertisers. Imagine our slogan—'Watch KFAD, the only station with a journalist on the air.' It could be big!"

"Well," said the station manager thoughtfully, "we did have one journalist who applied for a job a couple weeks ago. Claimed to have won something called a Pulitzer. I'm afraid I was rather rude to him. It lowers our image just to have them around."

"We're desperate, Boss, let's give it a try."

The station manager leaned back in his overstuffed chair, pushed away from his eight-foot conference table, lit an oversized cigar, and thought a moment. "A journalist . . . On TV . . . It's radical enough . . . It's even revolutionary!"

Then, he turned on his personal TV to watch a rerun of "LaVerne & Shirley."

SEPTEMBER 1993

Blood on Their Lenses

"If it bleeds, it leads" is local TV's aphorism which dictates its belief that violent crimes and traffic accidents, not the economy, lead off the nightly newscast. Focusing upon broken body parts is more "visual" and easier to cover than the economy, Senate hearings, and the health care crisis. And if it isn't violence that grabs the viewer, it can usually be something so soft that no self-respecting newspaper would run it any higher than page 17.

But, now and then, it's hard to find an assortment of accidents, fires, and murders. And so it was that KFAD-TV's panicked station manager met with his news director late one afternoon to go over the final line-up for the 6 O'Clock Aren't-We-All-Happy?-News. The station manager wasn't happy.

"What do you mean leading off the news with a report that some jokers at the Public Health Service found the cure for AIDS! Weren't there any accidents? Fires? Murders!"

"Sorry, Boss, there's nothing out there."

"Nothing!? Is that 'Nothing' as in 'no accidents,' or 'nothing' as in 'I'm close to finding another job in the industry'?!"

"Boss, we really tried. I've got five camera crews running around right now."

"Think you can get two of them to run into each other? We'd pay the hospital bills."

"Boss, don't you remember? The union made us agree to a six-month moratorium on stories that involve us maiming our crews just for the sake of ratings?"

"Some union," the station manager huffed. "Doesn't even want its members to get more air time."

"It's only for six months," said the news director. "After that, maybe we could cut the brake linings on Unit 3 and have Unit 4 cover it. But for now, the news scanner is dead."

"What happened to that fatality on Mulberry?"

"By the time we scrambled the chopper, the drivers had exchanged insurance numbers and left."

"Left!?" thundered the station manager. "No one leaves when there's a camera crew on the way!"

"Best we could figure out, it was just a few paint scratches."

"Any of the cars red? You guys get there faster and maybe it'd look like blood. Check the cops again."

"Sorry, Boss. Even Philly's not reporting any murders in the past 24 hours."

"Then go out and shoot someone!" the station manager demanded.

"Sorry, Boss, I can't do that."

"Yeah, you're right," said the station manager. "Tell Susie Sweetwater to do it. Her ratings are down. This oughta help."

"Susie's in the middle of her reading class right now, and you know how she hates to be disturbed when she's learning new words."

"Then Heartthrob! He's got the highest TVQ of any anchor in the country. Audiences salivate whenever he's on. The public would back him even if he had assault weapons and made welsh rarebit out of the Easter Bunny."

"It's an hour until air," the news director reminded the station manager. "Hearthrob's already in Makeup. They're darkening his hair tonight." "Roseanne!" shouted the station manager. "She's always good for something. Think we can get her to kill someone?"

"We have two crews on her now," said the news director, "but all she's doing is threatening a couple of writers and a producer or two. Besides, we've done that story 23 times this month."

"Check with the crew you have permanently assigned to Madonna. She's been quiet lately."

"Out of the country."

"Get me a fire! Forest. Trailer. Stove. I don't care what kind!" the station manager demanded, smashing his coffee mug against his desk, and cutting his wrist. "Blood!" he shouted. "We have blood!"

"It's only a scratch," said the news director.

"It's blood! And it's good for a grabber. Grab a producer. Come in with an extreme close-up full-frame, then pull back to a medium shot. Dissolve to some of the footage of the O.J. crime scene. Here's your lead: Violence in California leads to national blood-letting." He paused a moment. "Make sure you run teasers on this every five minutes."

AUGUST 1994

A Major Conflict of Interest

Like reporters everywhere, Keith Martin wanted to be where the action was, and during the first part of the year the action was in the Persian Gulf. Unlike the other reporters, Martin was a double agent. In fact, the other reporters at WBRE-TV, an NBC affiliate in Scranton/Wilkes-Barre, Pennsylvania, where Martin co-anchors the evening news, even proudly proclaimed his conflicting assignments—although they never mentioned the phrase "double agent." They didn't see it as a conflict.

Martin has been a TV journalist for more than two decades. He's also in the Pennsylvania National Guard. Not a grunt, but an officer. A *major*. As in management. And, he isn't just *any* officer, but a *public affairs officer*. Commander of the 109th Public Affairs Detachment. The military's very own flack.

Martin spent a week in February in the Persian Gulf reporting on the war and local units from northeastern Pennsylvania. He wasn't activated by the Department of Defense—apparently even the military have their limits on how many PAOs it can tolerate in combat zones. No, he was there as a journalist, although the distinction between flack and journalist blurred when on-air promotions and fellow reporters identified him as *Major* Martin. In addition to daily reports, Martin produced a one-hour special in which he interviewed eighty-five troops from Northeastern Pennsylvania.

Now with the war over, there are still stories to report. So, Maj. Keith Martin went to Fort Drum, New York, to report on the encampment of the 109th Infantry, a unit of the Pennsylvania National Guard. For the 6 p.m. newscast, he reported on the training the unit was getting so it can convert to being an armor unit by the end of the year. For the 11 p.m. newscast, he reported on the "excellent" safety record of the 109th.

Because Maj. Martin was on active duty at the time, a part of his 15-day a year commitment, he fed video and sound to all three stations in the Scranton/Wilkes-Barre market. Not surprisingly, only WBRE-TV aired his report. And, of course, he was identified on air as *Major* Martin. But that wasn't completely necessary since he was dressed, on air, in battle fatigues, complete with combat face paint.

The codes of ethics of the various journalism organizations are fairly clear about conflicts of interest. The Society of Professional Journalists states that journalists "must be free of obligation to any interest other than the public's right to know the truth. . . . Secondary employment . . . should be avoided if it compromises the integrity of journalists and their employers."

The Radio Television News Directors Association code states that "Broadcast journalists shall govern their . . . nonprofessional associations as may impinge on their professional activities in a manner that will protect them from conflict of interest, real or apparent."

The American Society of Newspaper Editors states that journalists "must avoid . . . any conflict of interest or the appearance of conflict." And the Associated Press Managing Editors code declares that journalists "should make every effort to be free of obligations to news sources and special interests . . . Outside employment that conflicts with news interests should be avoided. Secondary employment by news sources is an obvious conflict."

So, what does Larry Stirewald, WBRE-TV news director, say about all this? Stirewald doesn't believe there is a conflict of interest. "We make it very clear that Keith is in the National Guard," says Stirewald, noting that "as long as you're straight forward with the people, it's all right." He says that all journalists have conflicts, and that "being a professional means compensating for whatever baggage you're carrying." Then he asks the question, "Why *shouldn't* journalists [be able to] serve their country in the National Guard?"

Journalists, if they choose, *should* be able to be in the Guard or any of the military Reserves. Or even the Rotary Club. They just shouldn't be reporting about them.

SEPTEMBER 1992

The Intruder

He came to the front door late one Saturday morning and knocked. When there was no answer, he knocked again, then began yelling "HELLO," hoping someone would answer him. When no one answered, he opened the front door, and walked into the house.

"I heard a man shouting 'Hello,' says Rachel Kerr, a 21-year-old speech pathology major at Bloomsburg University of Pennsylvania, and one of two women in the sorority house at the time, "but I didn't go down because I thought it was for someone else."

He walked up the first flight of stairs of the house. Kerr, in her bedroom at the time, heard him, but didn't know if he was friend, repairman, or burglar.

"HELLO!" he again shouted. No answer. So, he walked down to the living room, looked around, left, then returned to the other side of the house and looked around. At the top of the stairwell, alerted by a sorority sister who had just come into the house, was Michelle May, sorority president and a 21-year-old speech communications major.

Before he could identify himself as a reporter for WNEP-TV, Wilkes-Barre/Scranton, May sharply told him, "I know who you are. We have nothing to say to you."

He politely asked if there was anyone he could talk with. "I'm the president," said May, "and we don't want to talk with you."

The reporter probably thought he was being fair, trying to get the other side of the story—or at least a decent 10-second "sound bite." After all, a 19-year-old pledge nearly died from alcohol overdose two days earlier in what was rumored to be a hazing incident.

"Can you at least tell me her name?" he asked. May refused, then told him, "You can walk out of this house, just like you walked in." He walked out of the house.

The reporter says his station's policy is that its reporters may not talk to the media. Officially, he has "no comment."

The women of Chi Sigma Rho were frightened, upset, and acknowledge they may not have been as courteous as they could have been. "We felt violated," said one of the women.

The reporter trespassed upon private property, according to Bob Buehner Jr., Montour County district attorney. The entry, he says, falls under the category of intentional tort, a civil action. However, Buehner also says there probably was nothing criminal about the reporter's entry, and any charges would probably be dismissed in court if there was no damage. For there to have been a defiant trespass, says Buehner, the reporter had to be in the house after being told to leave. To be charged with burglary or breaking and entering, he had to be in the house with the purpose of committing a crime.

Nevertheless, the reporter's actions, although not criminal, violated the ethics and standards of the journalism profession.

"You just don't enter someone's house without being invited," says Reggie Stuart, assistant news editor for the Knight-Ridder News Service and president of the 15,000-member Society of

Professional Journalists (SPJ). "If you do walk in," he says, "you should stay in the living room until you're recognized. You should never, under any circumstance, proceed to go through the house."

David Bartlett, president of the Radio-Television News Directors Association (RTNDA), agrees. "I don't think I would have gone into the house," he says, noting the reporter probably didn't exercise "very good judgment." Item 3 of the RTNDA Code of Ethics calls for reporters to "Respect the dignity, privacy, and well-being of people with whom they deal." Bartlett says that the situation in which a reporter enters a house uninvited "might be construed as an invasion of their dignity." The SPJ Code of Ethics specifically states that not only must the media "guard against invading a person's right to privacy," but that "Journalists at all times will show respect for the dignity, privacy, rights, and well-being of people encountered in the course of gathering and presenting the news."

There is a universal implied consent for reporters to go onto private property when there is breaking news, such as a fire. Reporters also usually have access to privately-owned quasi-public institutions, such as malls. However, none of the codes recognize entering a house by opening the front door unless first invited.

During the next three days, the local media ran innumerable stories about the alcohol incident, many of the stories factually inaccurate, some based upon the WNEP-TV reporting.

Four days after the media first reported the alcohol incident, a WYOU-TV reporter and his cameraman stood on the sidewalk before the house and, say witnesses, "taunted" sorority members following a university-wide meeting. "This is your chance to tell your side of the story," he said, a perfectly acceptable request, and important in the issue of fairness. But, say witnesses, he also kept taunting the women. "Tell us. Tell us," he kept saying over and over. "You look guilty if you don't say anything. Why don't you talk? This is your chance." He even told the women, "You need to take advantage of the media for your own good." Three times the women asked the reporter to leave. That night, he identified Chi Sigma Rho as "the bad girl sorority," while the station aired footage of the women persistently yelling for the reporter to quit bothering them and to leave.

The public has a love-hate relationship with the media. They want the press to meet its obligations to make sure the public knows what's happening in society. But, they also hate the way the media go about their jobs. Surveys show media credibility is at one of its lowest levels historically. The public is upset not only with factual, grammatical, and spelling errors, but also with the media's failure to admit mistakes. They are also upset with poor writing and editing, reliance upon unnamed sources, glorification of the criminal and of the bizarre. They don't like the media's failure to report on itself as well as it tries to cover other American institutions, and perceived unfairness, political bias, conflicts of interest. But most of all, they don't like ambush journalism" and the invasion of individual rights and privacies.

"We know we were wrong to allow a minor to drink," says May. But, she and her sorority sisters also have every right to be upset with what the media may have done to them and to their own credibility.

MAY 1995

TV News 101: Intro to Makeup

It was the first of 114 episodes of the critically-acclaimed TV series, "Lou Grant," and I eagerly awaited seeing it again, now in syndicated re-runs. The first episode reintroduced Ed Asner to American television as the crotchety, hard-driving, but cuddly journalist who once worked on a Detroit newspaper and had just been fired as news director of WJM-TV, Minneapolis—you know, the one where Mary Richards was a producer, and Murray Slaughter (before becoming the "Love Boat" captain) wrote copy for the bumbling Ted Baxter who was everyone's idea of the typical anchor. Lou was 50 years old and had $280 in the bank.

Coming to his rescue was Charlie Hume, a once-tough reporter now buried as a corporate-leaning managing editor, who said Lou would be perfect as city editor of the *L.A. Tribune*—as long as Mrs. Pynchon, the publisher, agrees. But, just one word of caution, said the editor—when meeting the publisher, avoid mentioning those years on TV—"she hates it."

"Then what do I tell her I've been doing the last ten years?" Lou asked.

"Tell her you were in jail," said Charlie.

And so it began. The first episode was directed by Gene Reynolds of *M*A*S*H* fame, from a brilliant script by Leon Tokatyan, but some of Hollywood's finest directors and writers worked on the series. In each episode were social and journalistic issues that made the one-hour show the best journalism classroom any student could imagine. But, I always remembered those early throw-a-way lines of cynicism between Lou and Charlie, that it was better to admit to having been in jail than to having worked in TV news.

The more I watch local TV news, the more I am convinced that even the worst of our local newspapers—and, certainly, the print media seem to have as many problems as there are lines of type—may be better than the average local newscast.

The typical TV newsroom has about as many reporters as a small daily, but at least ten times the news to cover. To compensate for the lack of reporters, the station leases a helicopter to give it "a presence in the market," hires three meteorologists to tell us the weather in Muscogee, Oklahoma, and two or three washed-up or never-was athletes to pump the latest exploits of the local Wattabago Whales Single-A baseball team. But, the station doesn't give us stories about the environment, health, labor, or the economy. What it does give us is gore.

To lead off the news, local stations send what few reporters they have to fires and car crashes to get that "visual" and to shove microphones up someone's nose and ask, "How do you feel now that your house burned down and your only son died in the inferno?" The rush to "quickie news" now dominates the local newscasts. "The locals use any old barn buring or jack-knifed trailer truck, and pass that kind of thing off as news," former CBS anchor Walter Cronkite told *Women's Day* readers in 1995. Station executives, he says, "have probably costed it out and that even maintaining the satellite trucks to run around to crime scenes is cheaper than having enough people on staff who can understand the news and spend time reporting it."

Find good reporters—and TV news does have good reporters—and their station management limits them to about 90 seconds of air time, no more than about three paragraphs in the local newspaper, then restricts their budget to nothing more than covering the local Miss Petunia Pageant, while encouraging the sports reporters to chase college and pro teams all over the country.

Unable—or unwilling—to dig out stories of substance, the typical 30-minute newscast gives us eight minutes of commercials, five minutes of sports, four minutes of weather, a couple of minutes of nauseating happy talk banter broken up into 10-second bites, a minute or two of canned network cast-off news, a minute of teasers and "upcoming at 11" self-promotion, and no more than ten minutes of local news. That, of course, leaves no time for local editorials, which is fine with management since it believes cordial relations with viewers and advertisers is preferred to journalistic integrity. In fact, the official CBS-TV policy is a very stark, "We do not take advocacy positions on controversial issues."

Behind all this are the ubiquitous "consultants" who are paid fortunes to tell station managers that stories about some starlet "accidentally" losing her bra in the surf is more substance than how the latest Senate hearings on health care reform will impact the local market, and that the station could squeeze another one-tenth of a rating point by lighting the set differently to highlight the anchor's "pool green" eyes.

It's not as if local TV doesn't have any models. "60 Minutes," "Nightline," and just about anything on CNN can go one-on-one with the best newspapers and news magazines.

Charles Kuralt, one of the nation's most distinguished network correspondents, once explained what TV News has become: "Urgent electronic music plays, the lights come up, and an earnest young man or woman says to a camera, 'Good evening, here is the news.' This is said very urgently and with the appearance of sincerity—most often by an attractive young person who would not know a news story if it jumped up and mussed his coiffure."

Andy Lack, NBC News president, is even harsher. "There's a generation of reporters coming along who are more interested in who their agent is or whether they're 'in play' at Fox than in going after the big stories," he charges. Because many of the reporters aren't willing "to invest in the homework that comes with really developing your skills as a first-rate reporter," says Lack, "there's very little emphasis on quality of work and a great deal of emphasis on compensation for it."

CBS anchor Dan Rather simply calls much of what passes as TV News "fuzz and wuzz."

Maybe if the journalists, both print and broadcast, learned more about their profession by watching what happened at the fictional *L.A. Tribune*, then maybe Lou Grant could have admitted he once worked in TV News, and we wouldn't have thousands of journalism students begging for a chance to be the next million dollar anchor, and preparing for it by checking hair styles instead of courthouse records.

DECEMBER 1994

Politically Incorrect Weather

A high-pressure front swept across the newsroom of KMJ-AM, Fresno, California, with a 100 percent certainty that it would leave a new low in its wake.

Depending upon whom you believe, weatherman Sean Boyd was either fired for being accurate, or program director John Broeske had finally had enough of Boyd's insolence and refusal to be "a team player." Boyd, an independent contractor, had forecast weather as many as 20 times a day for 17 years at the station.

However, for at least the second time in two months, Boyd was politically incorrect, a sin for anyone in our society, but especially so for a weatherman whose continued employment may have been based upon being meteorologically inaccurate.

In March 1995, Boyd had determined there was a better than average possibility that it would be partly cloudy, breezy, and cool for a station-sponsored golf tournament. Boyd says he remembers his program director suggesting that it would be better if the forecast could say it was partly sunny, not partly cloudy. Actually, why not report it would be mostly sunny with highs in the 70s and "let people make their own decision," Boyd remembers Broeske telling him. Nevertheless, Boyd stuck by his forecast; the weather for the divot swingers that afternoon was, in fact, partly cloudy, breezy, and cool.

Two months later, Boyd again got into trouble. KMJ, which carries the Rush Limbaugh talk-show, was sponsoring the Second Annual Dittohead Barbecue and Politically Incorrect Picnic, Saturday, April 15. Since it was already politically incorrect, it was apparently no big deal that it was also Easter and Passover weekend.

Four days before the Dittoheads were to meet, Boyd forecast a chance of showers, based upon reports of the National Weather Service. Broeske possibly didn't think that anyone, including the clouds, had a right to rain upon his plumped-up sizzling raucous barbecued hotdogs. After all, the station, which sends 5,000 screaming watts of conservative thinking into one of the state's biggest media markets, had something of an investment in making sure thousands showed up to celebrate the Biggest Mouth That Roars, even if the Chief Dittohead himself had no plans to attend. So, Broeske strongly suggested that Boyd revise the forecast. Afterall, you never know with California weather. It *could* have been a wonderful day in the neighborhood.

"Do you want me to change all the forecasts?" Boyd asked, then sarcastically suggested that the program director could just write down what *he* thought the weather should be, "and I'll tell it just the way you have asked." The program director might have thought about the temptation to move a few clouds but declined. Boyd stuck with his weather report. About 3 p.m. that Sunday, rain began falling upon the char-smoked Dittoheads.

Boyd says he and Broeske had "meetings in the past about how he wanted me to do things." The station manager, says Boyd, even told him that the program director had once mentioned that dealing with Boyd was "like a Chinese water torture."

Ten days after the Dittohead Debacle, Broeske approached Boyd. The Arctic Clipper cold front came with more warmth. "You can say you resigned because of stress and long hours," Boyd remembers Broeske telling him.

Broeske has a different version of the events. "Everything he's saying is completely untrue," says Broeske. However, he cites "station policy" for reasons why he won't discuss the events further. "No comment" is the extent of Broeske's official version.

Apparently, other media and their listeners don't have a problem with Boyd's personality or forecasts. Since 1988, Boyd has provided weather information throughout the day, including weekends when necessary, for the independently-owned KAAT-AM/KTNS-FM stations in Oakhurst. "Sean has been very responsible to us and to our listeners," says Larry Gamble, station manager and owner of the radio stations. He says that Boyd not only "provides the detailed information necessary," but also receives "very positive feedback from our listeners." Without question, Gamble is "very pleased" with the quality of the forecasts.

At KSEE-TV, an NBC affiliate in Fresno, news director Eric Hulnick also has no question about Boyd's competence. In the eight TV-station metro market, Boyd not only "has the most accurate forecasts in the Valley," says Hulnick, he's quite simply "the best forecaster in Fresno."

The best forecaster, unfortunately, now has a much smaller audience.

MAY 1995

Reel Violence

It was yet another stop on the book promotion trail, this time in Philadelphia on a "big-time" talk show with a "big-name" star. The host was friendly, and discussed my background and the book, a history of animated cartoons, although like most hosts she hadn't had a chance to read any of it.

"Let's get started by finding out what your favorite cartoon show is," she asked.

"I'm partial to the Roadrunner and Coyote series," I said, then briefly explained how the cartoons, with brilliant writing by Mike Maltese and directing by Chuck Jones, were classic throwbacks to some of the best silent physical comedies of the 1910s and 1920s. I expected an equally soft follow-up question. It came loaded with an explosive not even the Acme Co., the Coyote's supplier, could produce.

"There really is too much violence in cartoons, isn't there?" she rhetorically stated, then spent two minutes explaining her views.

"Actually," I said when she finally had to breathe, "the physical violence in cartoons is completely different from what you see in live-action or even in cartoons with human subjects." I got a couple of more sentences in when she came back, expounding the belief that cartoon violence directly leads to violence in real life, and that the studios and networks needed to be more responsible. Perhaps the Industry should establish a commission to review films or cartoons, she suggested.

Keeping my composure, I politely explained that the basis of all literature is conflict, and that most three-year-olds know the difference between cartoon violence and "real" violence, and if they don't, then parents should learn how to change the channels. Later, I was able to sneak in my opinion that it was absurd when network television, scared by lobbyists, had temporarily pulled Bugs Bunny cartoons from the air because they didn't think Elmer  Fudd should be blasting rabbits and ducks. She came right back at me by pretentiously quoting a research study to support her views, took a triumphant breath, and awaited what she thought would be my feeble response. Fifteen minutes into what I thought was a mugging—I had wanted to talk about bunnies and tweety birds—I fired back. "I'm well aware of that study," I said, then began to cite other studies that revealed either a slightly negative correlation or no correlation at all between cartoon violence and human action.

"Let's go to the phones," she said. For the most part, the audience asked interesting questions, with the host usually spending more time in presenting her views than I did in answering audience questions. Then, abruptly, she mellowed. "You certainly have a wealth of knowledge," she cooed. "I was wondering, do you have a favorite cartoon show?" Apparently, since I didn't answer correctly the first time, I got another chance.

"Rocky and Bullwinkle," I replied, explaining that Jay Ward's creation probably had the sharpest satire of all television shows. I was going to elaborate when she again explained that the plotting done by Boris and Natasha to the Moose and Squirrel couldn't be very healthy for impressionable minds.

"I believe that some studies show that cartoons may affect persons already prone to violence," I said, "but have no effect on persons who are not themselves violent." Commercials saved me from her response.

Back on air, she again introduced me and cited the book I was huckstering. "Let's go to the phones," she said again, and again the audience was more interested in the origin of cartoons, and some insight into the making of them. Five minutes before the hour, it was time to close it up, but not before one more question.

"By the way, one other thing before you leave," she asked, "what's your favorite cartoon show?"

This time I was determined to get it right. "Beany and Cecil?" I asked. When she said nothing, I briefly discussed the 1950s cartoon show created by Bob Clampett who had been one of the Warner Bros. pioneer directors. "I just loved all the puns and double entendres," I said, awaiting her response that cartoons were responsible for the moral breakdown of the American family, and that the world was at risk because of the conflict between Dishonest John and his targets Beany Boy and Cecil the Seasick Sea Serpent. But, she didn't. All she said was, "That's nice," thanked me for showing up, again mentioned the book, and went to another set of commercials.

I left the studio convinced I was yet another batch of chum for talk-show sharks—and wondering if I would ever get my favorite cartoon show right.

JULY 1995

An American Triangle: Violence, The Media, and Individual Responsibility

In Ohio, a five-year-old boy watches moronic TV cartoon characters Beavis and Butt-head, imitates their reckless use of matches, burns down the family trailer and kills his two-year-old sister. The mother, who had left the children alone, the fire department, and most of the community blame the cartoon. MTV eliminates the pyromania scenes from cartoons.

In Pennsylvania, Texas, New Jersey, and New York, five teenagers, imitating a scene from "The Program," a film about excessive football conditioning, lie down on the median lines of highways. Three are killed; two are critically injured. The communities blame the film, and Touchstone Films responds by deleting the scene from all prints.

In California, a 13-year-old boy murders a friend's father, then pours salt into the knife wounds. "I just seen it on TV," he confesses.

In Florida, a 15-year-old boy murders a neighbor. In Rhode Island, a boy hangs himself after watching a magician perform a similar stunt. The parents of both boys sue the TV networks for negligence.

And now in Washington, D.C., a media-rich city with one of the nation's highest crime rates, Congress is once again holding hearings on TV violence, and U. S. attorney general Janet Reno threatens the TV networks to either reduce the levels of on-air violence or face governmental intervention. She has a large base of support.

During the past four decades, there have been about 3,000 media-and-violence studies. No one disputes the facts that in the four decades that television has become a major form of recreation, society has become more violent. The National Coalition on Television Violence claims about one-fourth of all prime-time TV shows are "violent." The average child watches 27 hours of television a week, with the TV babysitting in the average inner city urban home about 70 hours a week. By high school graduation, the average child, according to some estimates, will have been exposed to about 12,000 TV murders and about 200,000 acts of violence on television.

Even if we eliminate the preposterous assertion by some groups that the Roadrunner-Coyote cartoons are among the most violent shows on the air, the reality is that there has been a significant increase of on-air violence. A few months ago, a Times-Mirror poll revealed that 72 percent of all Americans think TV entertainment shows contain too much violence. But, what people say—and what they watch—aren't comparable since the only statistics Hollywood notices are the box office or advertising revenue.

Since the media do not lead society, and usually pander to society's worst tastes, it's only logical to assume that American society has not only allowed—but by their viewing patterns encouraged—the increased levels of on-air violence. Without question, it seems as if some TV shows and many films are produced solely to package the most creative acts of violence to manipulate people into spending a couple of hours of escapist fun watching the conflict between good guys and bad guys kicking, biting, stomping, maiming, zapping, knifing, shooting, chopping, and crashing into each other.

Nevertheless, using the same data from the 3,000 studies, isn't it possible to conclude there is only a minimal cause-effect relationship? Isn't it possible that persons who normally have aggressive behavior patterns or who live in homes where such patterns are present may be more inclined to increase their own aggressive behaviors? And, isn't it possible that those without such personality traits or environmental exposure would not commit acts of violence even if *every* TV show had excessive levels? Certainly, it would be a leap of faith to believe that Quakers, if shown enough television violence, would buy guns and mug Baptists.

If we eliminated all media violence, as demanded by the shrill cries from our moral protectors, then we will be forced to eliminate most Nintendo and Genesis games, ban TV reruns of the "Three Stooges" and "M*A*S*H," pull the plug on MTV, smash tapes and CDs, stop singing "The Battle Hymn of the Republic," shelve the four-star four-hour film productions of *Gone With the Wind* and *Gettysburg*, stop the reprint publications of *Moby Dick*, *The Red Badge of Courage*, and *Uncle Tom's Cabin*, block publication of *Hansel and Gretel*, *Snow White*, and most of Grimm's fairy tales, rewrite virtually all of Shakespeare's plays, forbid the telling of Aesop's fables, prohibit TV news from appealing to our sense of the morbid by leading newscasts with auto crashes and wars, and order a complete rewrite of the Bible, which has more violent acts per page than even a Dirty Harry movie.

But, the federal government says it has no plans to curtail anything but the broadcast media. Something about the First Amendment. Besides, the government says it has the authority of the Communications Act of 1934 which declared that because of the limited number of airwaves, the government has the right to regulate radio and television to meet "the public interest, convenience and necessity." And, thus, we have a double standard. But, since 1934, we have developed newer technology that has created an unlimited information superhighway to allow the average TV set owner to receive more than 50 different signals. Within two years, home TV sets may be receiving more than 500 signals. If we continue to regulate content in one medium, no matter for whatever reason, and if we put even more fear into already greedy TV executives who will do whatever the government says as long as it doesn't pull their lucrative station licenses, then government will intrude upon all other media. Even if *all* TV shows become "violence-free," there will still be violence in American society, more than in any other civilized nation.

If the federal government truly wants to reduce violence in society, it might try looking at broader causes than the media. It might face up to the National Rifle Association lobby and severely restrict access to assault rifles and handguns. It might work to develop programs to reduce racism, poverty, war, worker exploitation, and unemployment. And for those persons who truly are violent, perhaps there needs to be tougher laws, better prisons, and more responsible rehabilitation programs.

Even more important, maybe we can start taking responsibility for our own actions and stop blaming the media, no matter how irresponsible, inept, or manipulative we think they may be.

OCTOBER 1993

Escalating an American Paranoia

Worried that a gang of thieves will break into your house, and that your dinky 9-mm. won't have enough firepower to stop them from taking your family's jewels? What if an alien nation launched an invasion, and your .357 Magnum liquified in your hands? How about those tank-sized rats in the basement who frightened your pet lion—or the moles who are building a city in your back yard and are planning to vote you out of office?

Fortunately, there's a solution. Get Rhino-Ammo, the "defensive" hyper-destructive hollow-nose bullets that fragment into hundreds of razorlike pieces within the body.

"The beauty behind it is that it makes an incredible wound," David Keen, owner of the Florida-based Signature Products, told the media the last week of 1994. "They're going to die," he said. "There's no way to stop the bleeding. I don't care where it hits. They're going down for good." The package for Rhino-Ammo proclaims, "the wound channel is catastrophic . . . Death is nearly instantaneous."

A second bullet, the Black Rhino, even more destructive, was designed to penetrate body armor, the kind police officers and some criminals wear. To allay Americans' fears, Keen promised that the Black Rhino would be sold "only to the right people," the 400,000 law enforcement officers and the 275,000 federally-licensed gun dealers. If the wrong people get the bullets, it's only because the "right" people gave it to them, said Keen. In a nation in which 1.1 million violent crimes were committed with a gun last year, and there were only 80,000 instances of the use of guns for self-defense, according to the Department of Justice, it's reassuring that only the "good guys" will get the bullets.

Federal laws prohibit the manufacture of "cop killer bullets," made of special metals, plastics, or coated with teflon. However, these bullets don't fall under these regulations because Keen claims they're not made of the prohibited materials.

The story of the Black Rhino began when Keen first fed it to his hometown newspaper, the *Huntsville Times*. After several other publications, including the *National Enquirer*, turned it down, Keen sent samples and test data to *Newsweek* when it expressed an interest in the story. The news brief in the December 19, 1994, issue opened with a description of what the bullets were likely to do, based upon Keen's claims on the package, then quoted an unnamed federal narcotics agent—"they'll be used against the good guys."

The Associated Press quickly picked up the story, leading its December 27 coverage with an ominous message—"Two bullets more deadly than those used in the long Island Rail Road massacre last year are about to be sold, despite fierce opposition from gun control advocates and police."

Within hours, most of the news media reported the latest "advance" in firepower, causing panic among the public, angry charges against Keen by the nation's police forces, and proposed federal legislation to ban "cop-killer" bullets, all reported by the national media.

The next day, Keen went on NBC-TV's "Today" show to declare that he was temporarily suspending production of the Black Rhino, but was continuing with plans to distribute the "less deadly" Rhino-ammo.

And then ABC's "Nightline" checked out Keen's claims, and determined that the Rhinos it tested were no more devastating than an average hollow-nose bullet, devastating enough but not "catastrophic." As for the Black Rhino, "Nightline" said it couldn't confirm, as of the end of the year, that such an armor-piercing bullet even existed. The news media now turned to claim the bullets were really a hoax.

Keen's response was that the bullet "Nightline" had was a "work-up" load Rhino, and not the actual bullets.

In its February 1995 issue, the *American Journalism Review* reported that Keen was "shooting blanks," and that the bullets were "too bad to be true." Keen says the *AJR* review was based solely upon other articles and not upon an analysis of the facts. "They slammed us," he says, bitterly claiming, "Nobody [in the media] lied to us more than" *AJR* in getting the story.

But, Keen faced attacks from an even deadlier foe. The National Rifle Association (NRA), knowing the public was irritable at widespread violence in America, had claimed, almost from the day of the Rhino's nation-wide disclosure, that the bullets were a hoax, certainly no more devastating than anything on the market. Further, the NRA also put the word out that Keen himself was a member—or at least a plant—of Handgun Control, Inc., whose purpose was meant to stir public fears about guns so they would support handgun registration.

"The only friendship I'm going to have is the NRA," Keen said, "and I've made an enemy that's very dangerous." Keen also claimed, "I am more frightened by the NRA than the death threats I have received."

Three months after the controversy began, Keen simply said, "It's been a very bizarre three months in my life." But it wasn't yet over.

In Spring, *Handgun* magazine conducted tests revealing that the .45 Rhino, now renamed the Razor, left a wound channel 5 inches wide and 7-1/2 inches long in a test against ordnance jelly. Jan Libourel, editor of the 180,000-circulation magazine, says the bullets, although not as catastrophic as what Keen claimed, were nevertheless more deadly than almost all other bullets on the market. However, critics point out that ordnance jelly, while often used to simulate humans, still is not human, and that not only was the magazine the "lapdog of the manufacturers," the magazine's "political agenda" was suspect.

Then in August, eight months after its first report of the catastrophic effect of the Rhinos, the Associated Press distributed an article that tended to confirm the *Handgun* report. The AP had contracted with the H.P. White Laboratory to test the Razor against three other high-tech 9 mm. bullets. Both the Razor and Glaser bullets, each of them composed of lead pellets within a plastic-filled shell, left almost identical wounds when fired into ordnance jelly. All four bullets—Razor, Glaser, MagSafe, and Federal—were deadlier than whatever else was available to civilians, but none could penetrate body armor. However, the laboratory cautioned that because the Razor and Glaser bullets require high gun pressures, the risk "is that it could damage your gun or cause it to blow up."

In order to get a story, says Keen, the media "pit one group against the other," with the controversy partially stirred by "information merchants eager for sensational headlines," but who didn't check out the truth.

Maybe Keen's original claims were all hype by a company that was trying to switch in a post-cold war era from making protective coatings for the Stealth aircraft to a consumer-based industry. Maybe Keen really does have a secret bullet, that Black Rhino he took out of production at the end of 1994, that is more devastating than anything else on the market. Whatever the truth is, the Rhino stirred America's fears, and the media were right there to tell us the bullet existed—or didn't exist—or that it may exist.

Either way, the story played well in headlines and the evening news for about two weeks, then died while other claims or hoaxes moved to the media's short attention span.

AUGUST 1995

Labels of Violence

On a hot, muggy summer day, with my brain waves set on pause, I decided it was time to check out the low prices at the nearby air-conditioned discount department store.

Because it's the school-less summer, the low prices everyday store made sure its toy section was fully stocked. Among Barbie dolls and Mutant Ninja Turtles were two aisles of fake guns, everything from 98-cent plastic water pistols to a Super Soaker 100 "for the serious watergun enthusiast."

Cap pistols are as popular today as when they were first invented in the late nineteenth century. For a buck ninety-seven, any kid can own a black 12-shot ring cap gun; extra caps are only 87 cents. If that model doesn't appeal to everyone, there's a dozen more that might, and none cost more than five bucks.

For those who want to play "cops 'n' robbers," for under four bucks each are copies of 9-mm. auto-power chargers pistols, an assortment of .22 short-barrel Saturday night specials, a clip-loaded dart gun, and a silver-plated die-cast metal repeater pistol that the package proclaims is "recommended for ages 4 and over." Of course, just to make sure the "good guys" always have the edge, there's an assortment of handcuffs, police batons, tin police badges, and Dick Tracy wrist-radios.

For the Old West historians among the bubble-gum set, there are bow-and-arrow sets and replicas of the Colt .45, advertised in newly-packaged die-cast metal as the "widow maker," and several "Old West" super-shot rifles that white men thought Indians called "sticks that pour fire," but which the Indians knew were the weapons of their own genocide.

Future GI Joes, inundated by non-stop Saturday morning TV watching, can pick from an assortment of almost every weapon of war except for full-scale Cruise missiles, and even these might be on some kindly old Gepetto's work bench almost ready for Christmas delivery.

For those with more immediate needs, there's water balloon grenades, X-1 recoil blasters, X-2 Nitro blasters "with a thunderous sound," Whistle Missile Skybolts, machine lasers, and a pulsating Fazer II with a vibrating handle and a box that begs children to "try me."

Especially popular among cookie crumblers are the high-tech "safe" Nerf guns, among them a Nerf sling shot, a "high powered, fast action," Nerf bow 'n' arrow, Nerf Missile Blaster, and the Nerf Missile Storm that can rapid fire four foam missiles. For the more affluent, there's the Nerf Hydro Bazooka with double punch on sale for just $14.83 and the Nerf Arrow Storm Gatling unit rapid fire system for a mere $22.96, the package larger than the three-year-old who's looking it over.

Just fifty feet away in the sporting goods section, among Bowie knives and an assortment of .22 single-shot rifles and 12-gauge and 20-gauge shotguns, are "fast shooting, accurate," BB pistols and rifles "with constant firepower," all for under $50 each.

Want even *more* firepower? Just go about a quarter-mile away to another store where a new Smith & Wesson .357 magnum is only $269.95. Can't afford that much? Pawn shops in any larger city have .22s for under a hundred dollars or, for under $500, semi-automatic military assault rifles that can easily be converted into illegal 900-shot-per-minute automatic weapons.

Last year, there were more than 60,000 deaths, about 2,000 of them accidental, from firearms. Another 90,000 were injured. Among high school students, gunshot wounds are the second leading cause of death. Of the 200 million firearms in the country, 60 million are handguns. Only the National Rifle Association believes it's a Constitutional right for individuals to bear arms because even the courts have repeatedly ruled the Second Amendment applies to militias, not individuals.

And now comes the television networks which have figured out if they didn't do anything about violence on the public airwaves, then Congress will. So, the geniuses who gave us "Cop Rock" and numerous variations of the Amy Fisher story, cancelling the critically-acclaimed "I'll Fly Away" and "Brooklyn Bridge," decided a label will help reduce violence in American society.

These same six-figure blow-dried TV geniuses—whose window-darkened limousines pick them up from fenced-in suburban estates and drive them to work in circuitous patterns to avoid seeing the city—are now trying to make us believe that labels, similar to the movies' rating system, are going to make our children less prone to imitate what they watch. And they sound like they actually care about God, mother, apple pie, and the right of all citizens to bear arms and still be free from violence.

Of course, we are carefully assured by the networks—whose cash crop on Saturday mornings is toy companies—that all the current TV shows are so peaceful and respectful of human life that none qualify for the label. It's the future shows that may need labels, they say. Exempt, of course, are a plethora of shows and the TV news which, reflecting American society, is probably the most violent half-hour on the air.

Nevertheless, we're going to have labels. Just like the labels that Tipper Gore got record producers to put on albums, and which teenagers now scavenge record stores to find. Just like the labels that tell 13-year-olds that "R"-rated movies are the ones with the gratuitous sex scenes, filthy language, and excessive violence. The same labels that are supposed to keep mothers from taking their seven-year-old daughters to see "Jurassic Park" which has a PG-13 label. Labels don't work for record albums; they don't work for the movies. And until we reduce violence in society, there's no way they're going to work for television.

JULY 1993

Compliments of a Thief

They won't tell you their names, but they'll sell you a genuine knock-off Rolex for only fifty bucks. Too high? How about forty? Thirty-five? But they can't go any lower; why, it's almost a *steal* at that price.

Don't want a watch? They have pretend 14-karat gold necklaces and rings? Something nice for that special lady? Still not interested? Wait! Don't go! Nice jeans? A bandanna?

It's the street vendors. They're in almost all major cities trying to make a buck. Most don't make a lot of money, just enough to survive, enough for a seedy but overpriced walk-up; some clothes; a decent meal once a day. Many are immigrants, here in urban America to find the "good life." Most don't know, at least not "for sure" know, if their merchandise, provided by middlemen, is stolen or just purchased in large quantities at fantastic wholesale prices.

Like the street vendors, a few "needy" college professors also have something to sell. However, they're the amateurs, and their buyers are the pros. At the end of every semester, book buyers descend upon the college campuses to buy books from the profs. Not the used books that students sell back to the bookstore the day after their finals, but new books. Complimentary ones. Books supplied by publishers, often at the professor's request. The purpose of complimentary copies, sometimes as many as 5,000 per press run, is to entice the nation's professors to adopt the books for a course.

No one knows how many of the nation's 600,000 professors sell comp copies, although good estimates are that of more than 100 million college texts published a year, as many as half of the estimated one million complimentary copies may make it into paid distribution.

Almost everyone benefits. The profs make out well since they can sometimes make $300-400 in undeclared income merely for opening mailing bags a few times a semester and occasionally thumbing through the merchandise.

The agents make out well since they buy $40 books for, maybe, $5. And the wholesalers and bookstores make out well since they buy books at far less than half the cost of new books.

Not making out so well are the publishers and authors. In 1986, the last time the Book Industry Study Group checked, authors were losing $10 million a year in royalties, and publishers were losing $80 million a year in sales to the comp book racket. A few years ago, Karl J. Smith, a math professor at Santa Rosa (California) Junior College, quickly learned how bad the problem was; more than half the students in his class had "used books"—although the book he wrote had just been released three weeks earlier.

Professors, in rebuttal, say that many of the copies are unsolicited, so it's their right to sell them. They say even books they asked for may not, after inspection, be appropriate for the courses they teach. The publishers suggest sending back the unused books, and often include postage-paid coupons and mailers; they suggest the professor may place the books on department library shelves or even donate the books to charitable agencies. But, the comp copies are still being sold for "spare change."

To stop the sale of complimentary copies, publishers have begun embossing "Complimentary—Not for Sale" on the covers, and stamped the same message on the end of the pages. But, wholesalers have placed non-removable "USED" stickers on the covers, and sanded off the message on the pages. Many wholesalers even rebind some titles.

Many colleges have written policies that forbid the professors to sell their comp copies. However, professors still find ways to circumvent the policies.

"We strongly suggest that stores don't buy comp copies," says Jerry Buchs of the 3,000-member National Association of College Stores. But, the Association can only recommend since it has no enforcement powers in its code of ethics. Nevertheless, says Buchs, "We keep addressing the issue."

Except for authors and publishers not receiving money for

writing and producing books, and some ethical considerations swirling around professors making money from books they had no part in creating, students wonder what the problem is. After all, they're getting new books at "used" prices. The problem is that the sale of complimentary textbooks to agents—not the sale of legitimately purchased used books the students sell back, but which also yield no income to authors and publishers—directly leads to higher list prices for all textbooks, says James Lichtenberg, vice-president of the Association of American Publishers (AAP). It's rare that a student doesn't complain about the high price of textbooks; they should be complaining about their greedy, unethical professors.

JULY 1995

Stripping Off Their Royalties

He's there by 7 a.m. almost every Sunday except in Winter to make money at one of the largest permanent flea markets in northeastern Pennsylvania. In three-foot long cardboard boxes he has an inventory of hundreds of paperbacks, all of them displayed spine up. Westerns. Romances. Adventures. Whatever you want. Three for a buck; fifty cents each. The books are virtually mint condition, and if you don't mind reading something without a front cover, it's a bargain, especially since paperbacks with the covers, sold at supermarkets, pharmacies, and bookstores, are now going for $5.95 each. The only problem is that it's illegal.

The sale of stripped books, says Roger Williams of the Association of American Publishers (AAP), is a "significant and ongoing problem" that involves fraud, possible copyright infringement, and often federal interstate commerce violations.

However, Eric Raymond, an executive at Simon & Schuster, one of the nation's largest publishing houses, says that police departments and prosecutors often don't have the time, manpower, or resources to investigate and bring to court sellers of stripped books. "It's just not the thing prosecutors want to spend time with," he says. Nevertheless, because of the volume of lost sales, publishers have begun their own investigations and have been working with both local and federal law enforcement officials. The FBI, as is its policy, refuses to discuss ongoing investigations.

To understand why the sale of stripped books is illegal, it's important to know a little about the nature of book publishing. Although the major book chains usually buy books on the basis of a book's cover and the promotion effort put out by the publisher, no one can predict which books will titillate American reading appetites. So, publishers of the mass market paperbacks—the kind with colorfully-embossed titles superimposed over pirates and scantily-clad women on slick 4-1/4 by 6-3/4 inch covers—order large print quantities to try to saturate American bookstands. They sell these books to distributors for 50-55 percent of the list price, and hope a few titles bring in enough profit to carry the rest of the line.

Unique in the field of retail sales, booksellers can return to publishers for full credit any books they can't sell. However, publishers have no desire to pay shipping costs for books they probably won't redistribute, especially since there are another couple of dozen titles they're trying to push that month. And, neither bookseller nor publisher wants several skids of taxable inventory. So, distributors and publishers sign contracts that allow the bookseller to send only the cover back to the publisher, tack on shipping costs, and agree to destroy the rest of the book to prevent further sale.

The bookseller usually sends stripped books to a recycler who picks them up at no cost and makes his money by selling recycled pulp. Last year, 728.5 million mass market paperbacks were published, but about 310 million of them were credited as unsold and recycled, according to the AAP.

However, some booksellers "forget" to send some books to a debindery or recycler, either selling some in their own store or, more likely, selling books for pennies apiece to mini-distributors. But, even if the bookseller (who can be the owner of just about any kind of a business) plays by all the rules—and almost all major bookstores do—and sends the books to a recycler, that doesn't mean the books don't show up again. Some books may be stolen in transit or in storage; and, a few unscrupulous companies may file claims they have shredded 10 tons of what is now literally literary garbage, but have really gotten rid of just nine tons, throwing the coverless books into the streets, like left over food for the cats. The cats, in this case, have pick-ups, and pay for the leftovers.

So, what's really the problem? After all, even though these transient booksellers probably don't pay taxes on their income, it's hard enough these days to make a buck. And, certainly, it's a break for the readers who are more likely to buy a 50-cent paperback than one costing ten times as much.

The problem is that when a reader buys a stripped paperback, the publisher doesn't receive any money. Since there's no income to the publisher, there's also no royalties to the author who is usually paid 5 percent of the list price of mass market paperbacks.

Except for the few million-dollar deals that make the headlines every now and then, we authors don't make a lot of money anyhow from our meager royalties,which average 10-15 percent of the list price of hardcover books and 5-10 percent of paperbacks. So every stripped or stolen book that's sold means we get no money while a lot of people who had no part in the creative process are making money off of us. And, I really object to that.
AUGUST 1994

The Hustle

The hustle had begun less than six hours after they buried the producer when the production manager on a 30-minute "infomercial" decided that he wanted four times more money since he figured it was now more difficult without the producer.

The widow, who had been in "The Industry" for 17 years, said there wasn't that much money, especially since the production manager had taken it upon himself that evening to hire another production assistant and a script clerk whose only talents seemed to be that she smiled a lot at the production manager.

But the production manager knew there was now no way production could be completed without him—and his friend. The friend was a salesman who somehow had become the star's agent, then decided to become a producer. After all, with his gold-color Cadillac, $150 shoes, open shirt, and three golden chains strung around his neck, he *looked* Hollywood.

In his best macho voice, the salesman-producer proclaimed to cast and crew, "We'll do whatever is fair." He decided it was "fair" for everyone, including him, to get more money. Naturally, the additional expense would come from the widow's pocket—and from the "star" who was convinced that "good" production needed even more money. Well, thought the widow, at least she could count on the Russians not doing anything stupid. She was wrong.

The "Russians" were two Russian immigrants who had been neophyte filmmakers in the Soviet Union, but who had defected because they were denied "artistic freedom." Neither was political; they just wanted a chance to get more work.

The now-deceased American producer-director had been the only one who had given the Russians film work in more than three years. When the producer-director died, the Russian first assistant director became the director at twice his original salary. Within hours, he claimed the American director of photography didn't know what he was doing, and gave control to the other Russian who had originally been hired as a production assistant.

The day after the producer was buried, the Russians demanded even more money. Naturally, the salesman-agent agreed; it was, as he proudly proclaimed, "only fair." Then, a young assistant cameraman decided that twice his contract rate was "only fair." And someone else submitted an expense voucher for eight hours driving time for a hundred mile trip.

Next, the salesman-agent who thought he was a producer decided that the script needed changing, so he changed it without consulting the writer or director, and set back production another two days, and several thousand more dollars. Then it was time to tell the film editor and sound engineer they weren't competent, and to dismiss both of them, although the film editor had taken numerous productions through final edit during his three decade career.

When the unions started questioning some work rule violations, the salesman-agent tried "sweet talking" them—he called it PR—and handed out some "gifts." Only because the now-deceased director-producer had such a good working relationship with the unions, and because of concern for the widow, did they return the gifts and decide not to shut down the production and bring the amateurs up on numerous charges, including stupidity.

So now, "The Magnificent Four"—the production manager, the director, the cinematographer, and the salesman-agent—were going to edit the film—if they could just get it out of the lab.

"The film won't cut," the widow had been saying all along, knowing that the way it was shot after her husband's death was so scattered that scenes didn't match, and that an editor would have an impossible time putting it all together.

"You're crazy!" they told her, and kept shooting "fill-in" scenes.

"You don't have a film," she now firmly told them, much to their sarcasm. But, when they tried to get the negative out of the film lab, they were told that, indeed, they didn't have a film. By law, the widow was the only one who could release the negative; after all, she was the one who had previously paid for it.

For the next six months, the four would-be filmmakers, their lawyers and cronies, tried every trick they could to get the film. None succeeded.

"Buy me out," she said cooly. They had no other choice. But at least they were now "Hollywood," and could use their production credit to convince other companies to invest in making infomericials, even if they didn't air except at 3 a.m.

MARCH 1994

The Write Stuff

It was close to deadline, and I was hurriedly flipping through newspapers and magazines, trying to find a news hook upon which to hang this week's column. Marshbaum was on vacation; I didn't have time to read the *Congressional Quarterly*, source of innumerable columns; and every two-bit humor "wannabe" was mining the Bar Association for cheap lawyer jokes.

Dejectedly, I thumbed the classified ads—maybe there'd be a job for a washed-up columnist. My mind wandered over to the "literary services" section. *There!* In *Harper's*, one of the nation's most respected magazines, was my salvation. Among ads from vanity publishing companies promising to make me a star was exactly what I needed. Three companies said they'd write this week's column for me.

The first ad promised, "We write everything. Reports, papers, company books." The ad even claimed the company was "professional." Alas, it had only a mail address, and I needed something in less than a day.

Next was a Los Angeles company with a "toll-free hotline." John, a polite young man eager to help, told me most of the previously-written termpapers in the company's catalogue were 6-20 pages, and the cost was only $7.50 per page. The company was even so concerned about its clients' finances that it never charged for more than 17 pages, and threw in the footnotes and bibliography for free.

"I'm on a deadline," I told him.

"No problem," he helpfully said, "we can FAX it to you or send it by overnight mail."

I told him I really didn't need any of the advertised 19,278 termpapers, but thought maybe there could be some "special" assistance he could provide. There was no problem there either. For only $20-$25 a page, the company would custom make an "undergraduate level" report to my specific needs. For $25-$45 per page, I could get a graduate or professional level report. It could be six pages; it could be 400 pages. My choice. "Of course," he said, "we expect you to put in some of your own opinions." Of course.

In exchange for this "fully written report," I'd have to sign a statement guaranteeing, "I understand this report is to be used for research purposes only." Right.

I opted for the professional level report, and was told to call another number and to talk with a Dr. Something-or-the-Other who was in charge of the researchers. Not wanting to lower the quality of my column, I first asked her about qualifications, and was told "every one" of the staff has at least a master's. I didn't have time to verify writing ability, subject knowledge, or even if the M.A.s were from the schools that advertised in the classifieds of slick magazines. Alas, she said the company doesn't do fiction. "Try a grad student at a university," she suggested. "Someone in English might be able to help you." A grad student? In English! Obviously, this woman had the artistic sensitivity of a drain pipe.

Time for my last contact. A kindly voice answered, "Research Services."

"Can you do a custom report?" I asked.

"Everything we do is custom made," he replied. He charged $12.50 per page, with a special rate of $200 for the first 15 pages. I'd have to supply the cover page, but he'd throw in footnotes and bibliography at no cost.

"Can you do satire?" I asked. No problem, he answered. However, he explained that writing satire "is time consuming because there's no library research, and it involves creative writing." I said I understood.

"I'd like a foil in it if possible. You know, like Mike Royko uses Slats Grobnik, and this writer in Pennsylvania uses a guy named Marshbaum?" Still no problem.

"I need it pretty soon," I begged. No problem there either. Except that with satire, he said that the more time he had to write the "report," the better since "if you do it too fast, a good idea may come after it's in the mail." But, for $50 he'd do something. Three pages. Even with a foil.

That's when I decided not to hire him. I spend all week perfecting my column. And he wanted only fifty dollars? For 800 choice words? It seemed awfully cheap. Besides, by then, I had my column.

DECEMBER 1993

Rerunning America

At one time, television was divided into 13-week blocks. This allowed news directors to fire anchors whose personality quotient and hair style withered, and network programming chiefs to dump a series whose ratings dropped faster than a politician's ethics. There were three 13-week blocks a year for entertainment shows; the fourth block was for summer reruns. However, the Suits-with-Gold-Chains figured it was not only less expensive to rerun a prime-time show, but that the public will tolerate anything as long as it's a hundred variations of the same thing. The result is that there are only about 22 new episodes of each show a year; the rest of the time is spent on reruns or "specials." Of course, this means less money for writers, directors, and actors who are paid on a per-episode basis, and more for executives who are paid no matter how insignificant their brains.

Since reruns are so profitable for the networks, maybe we can adopt the same philosophy during our own summers.

We could rerun the critically-acclaimed, low-rated, highly-entertaining, now-cancelled TV series, "I'll Fly Away," and keep running it until the entire country sees—and understands—a critical part of America's civil rights history.

Let's rerun the Bette Midler and Barbra Streisand concert tours. It doesn't get much better than that.

Maybe, we could rerun some things and actually improve upon them. We definitely must rerun the Winter Olympics coverage. This time, we lose the media hype on the Nancy and Tonya Smash-and-Cry Show, and focus upon the athletes who deserved, but never got, the coverage. And, while we're at it, let's also lose the hype surrounding Amy Fisher, Joey Buttafuco, the Menendez Murders, the Bobbitt Bisection, and anyone named Donald or Marla. (Maybe we could even rerun the California earthquake; this time, we put it under Trump Tower.)

We could definitely rerun this past winter, editing out both a few snow storms and all the TV meteorologists who kept cheering the snows on to an all-time seasonal record. Possibly, we could even take a few of the smaller snow storms we edited out of the winter's scenes and drop them into our current mucky, humid weather.

We could rerun the past session of the state legislature on wide-screen TVs, and hope we can finally spot intelligent life.

We could rerun Philadelphia's elections. This time, we rerun an honest election.

We must rerun the war in Bosnia before they're all casualties of international indifference. This time, we add a few scenes where the international community steps in and actually does something, rather than just talks about how brave those Bosnians are for enduring torture, rape, and the constant glow of Serbian artillery and TV camera lights.

We could even cancel the last couple of seasons of prime-time Somalia and Rwanda, thereby saving millions of lives.

Last year, hundreds of corporations laid off hundreds of thousands of workers. This Summer, we rerun these layoffs, but this time we lay off top management; after all, it wasn't the workers who caused the problems.

And, finally, there are an estimated four million instances of spousal abuse each year, according to the FBI. Husbands, former husbands, boyfriends, or former boyfriends account for about 1,400 murders of women a year, according to the Family Violence Research program of the University of Rhode Island. The Los Angeles Police report that on an average, about every nine days a murder occurs in L.A. from a domestic dispute. Let's rerun O.J. Simpson's domestic life. Maybe if the police had investigated, then arrested, Simpson the first time they had to be called out on a call of assault and battery . . . And, maybe, if he was found guilty, and if the Court ordered him to undergo a program of mental therapy, along with restrictions on his personal freedom . . . And, maybe, if he abused his wife again, he'd be ordered to jail and not released until society was assured he posed no threat to anyone . . . And, maybe, if the nation's media had concentrated upon the issues of domestic violence instead of what clothes the latest flock of starlets wore to the pre-Oscars, then, maybe, O.J. wouldn't be facing the death penalty today, and the nation wouldn't be wondering how such a nice man could have done such a violent act.

JUNE 1994

'Local Reporter Cops Lexicon Story'

About 24 centuries ago, Hippocrates said, "the chief virtue that language can have is clearness." It's too bad the great physician's words haven't yet made it into American media.

Almost every day, we read or hear about a "spectacular fire" or "spectacular accident" where (pick one) (a) a car, (b) a truck, or (c) a fleet of skateboards "careened out of control." Nonsense! Fireworks on the Fourth of July are spectacular. A "fatal fire" (reporters love alliteration) or a traffic accident in which three people are killed are *not* spectacular.

And, where else but in the media can we hear about some psycho with a high-power assault weapon holding "police at bay"? When the battle is over, we are likely to learn that the "perp"—who used to be a criminal before the debut of TV's *Hill Street Blues*—was "racked by guilt." And, since "racked" is such a popular piece of "journalese," we also learn there are people in our society, maybe even criminals just before their sentencing, who are also "racked by pain" and "racked by anguish." All this, of course, leaves most readers being "mental racks."

Has anyone else noticed that if a young lady is "brutally murdered" (as if some murders aren't brutal?) the lead often reads, "A pretty 27-year-old woman was killed late last night." Does anyone recall reading, "An ugly 27-year old was poisoned early this morning"? How about men who are killed? Do we identify them as "A 27-year-old Greek god hunk with rippling muscles was stabbed yesterday afternoon while pumping iron in his West Catcall basement"? So far, reporters haven't identified any murder victim as having been "drop dead gorgeous," a term used by the media to apply to numerous living actresses and models.

Should there be an "eyewitness," apparently someone more alert than just a plain, ordinary witness, we might learn that the criminal or victim was last seen "exiting the building at approximately 11 p.m." Apparently, "leaving the building about 11 p.m." doesn't sound as menacing.

To pretend we are covering all possible law suits, we drop in "alleged" now and then. Thus, if one "alleged" is good, then several of them apparently protect journalists from suits even more, as in "the alleged burglar was arrested by police who charged him with allegedly committing the alleged crime." One Denver TV reporter claimed "several alleged shots were fired," and a Bloomsburg, Pennsylvania, newspaper, apparently believing that everything must be attributed to someone, splashed a headline: "Stabbing victim dies, police say."

On coverage of labor issues, the media often report that both sides are "eyeball-to-eyeball," a condition that leads either to a "strike-bound condition" or a visit to an optometrist. Nevertheless, following these "tense moments," both sides "come to terms" and are now in the "process of hammering out a contract"—just as soon as they "nail down the clauses."

Sportswriters seem to believe language was invented so they could butcher it. Athletes no longer exist; they're booters and gridders, tankmen, grapplers, and hoopsters. As for the teams themselves, they don't defeat an opponent, they maul, exterminate, pulverize, crush, annihilate, and devastate them. Check out the headlines and see "cop" become a verb, as the "locals cop a victory." In baseball, pitchers no longer throw fastballs; they're flamethrowers who "toss the pill" or, even worse, "throw with velocity." A sports editor I once worked with wrote about a high school player who, with a few seconds left in the game, brought victory to his team when he "hooped the brown spheroid through the draped iron doughnut."

Of course, the language of the sports reporter is no match for the language of government which has 15-page recipes for a simple fruit cake, labels a parachute an "aerodynamic personnel decelerator," and tells us a toothpick is a "wooden interdental stimulator."

From the nation's 700,000 professors have come so much incoherent babbling disguised as professional papers, journal articles, and books that banning what poses as academic scholarship would help save the rain forests. Unfortunately, much of the bad writing comes from professors of mass communication, few of whom ever worked more than a few months in a newsroom, and who believe that conducting pseudoscientific research will give them respectability, or at least tenure. The following are just *parts* of titles of what passed as scholarship in one recent convention of an association of journalism professors: "symbolic modeling and persuasive efficacy information on self-efficacy beliefs," "heuristic perspectives," "contextualist cultural functionalism," "concept mapping," "conceptions of salience," "systemic methods of measuring free recall," "synchronous and asynchronous forums," "dyadic interaction," "news framing and audience framing," "structural pluralism," "discourse analysis," and "observations on a conundrum."

Indeed, most "mass communicologists" are more comfortable with running statistical analyses with two degrees of freedom, and in jabbering about channeling, cognitive dissonance, and content analysis than they are with leads, transitions, and conclusions. Perhaps, that's why we're still wondering why many recent journalism graduates can't write.

SEPTEMBER 1994

Linguistic Larceny

A recent governmental investigation of the Veterans Administration Hospital in Chicago revealed that many of the resident physicians were poorly supervised, and that there were substantial instances of crucial diagnostic mistakes and inappropriate surgery. Sometime while all this was going on, six patients died in the operating room, and the VA was forced to make some rather large cash payments to the families of those individuals. According to an official explanation, each of the deaths was a "surgical misadventure." Sort of like "Adventures in Paradise," or even "Pee-Wee's Big Adventure." Maybe the VA thought it was on a campout, and nothing worse occurred than the tent fell during the rain.

But, the VA isn't the only one to use the great linguistic cover-up. A few years ago, a Colorado state legislator proposed a plan that would clean up Denver's polluted atmosphere, at that time well within the nation's "Bottom 10." Instead of spending millions to clean up the air, the legislator suggested a little language manipulation would solve the problem. "Hazardous air" would become known as just "poor air." "Dangerous air" would be "acceptable," and "very unhealthful air" would be "fair air." As for just "unhealthful air," it would be "good air," so pure that smoke-puffing industries would just have to bottle it and sell it as distilled. Even the Environmental Protection Agency has helped make American air safer by simply reclassifying acid rain as "wet disposition."

Because truth is the first casualty of war, we willingly accept the great linguistic cover-up. Our Civil War wasn't a revolution, but a "war between the states." World War I, which we obtusely believed was "the war to end all wars," was the "Great War." Fortunately, no one labelled World War II the "greater war." At the beginning of the 1950s, we peace-loving Americans learned that a "police action" in Korea looked remarkably like a war and that peace was a "cold war."

A decade later, we were in an "undeclared war" known as the "Viet Nam Conflict" where "military advisors" helped direct "preventive air strikes." The use of Agent Orange to defoliate the countryside was merely employing a "resources control program," and we learned that to "exterminate with extreme prejudice" meant that someone was going to be killed. About this time, the Pentagon, feeling its linguistic superiority, tried to slip a "radiation enhancement device," a neutron bomb, past Congress.

It was during the Vietnam War that Gen. William Westmoreland, commanding about a half-million troops, declared that the reason the military wasn't giving the American people the truth was because "without censorship, things could get terribly confused in the public mind."

During the Persian Gulf War, we learned that "foul-ups" that caused injuries and death to our own troops were "accidents as the result of friendly fire." Our soldiers learned that exposure to nerve gas could result not in death but in "immediate permanent incapacitation." Warplanes were really "weapons systems" which were merely "visiting a site." If the pilots succeeded in "cleansing," "neutralizing," or "sanitizing" that site, they could report they "were successful in servicing their intended targets." If they over-bombed or missed entirely, wiping out civilians and their houses, the pilots merely reported a lot of "collateral damage."

Of course, we are well aware that many of the Arab Coalition allies believe that sand fleas are more important than women. But, did we really have to accept the Saudi Arabian demands not to include women in combat units then agree that among the half-million Coalition troops in Operation Desert Storm about 40,000 were "males with female features"?

"Post Traumatic Stress Syndrome"—known as "shell shock" in World War I, and "battle fatigue" in World War II—has become the universal way to explain actions not only of persons traumatized by war but also the actions of spouse-abusers and mass murderers as well, even if their only combat was changing typewriter ribbons in an air-conditioned HQ.

In Bosnia-Herzegovena, the Serbs aren't murdering, raping, and pillaging, they're merely on a campaign of "ethnic cleansing." When that war is over, the Serbs will probably declare themselves "victims of society," and will experience a lot of "post-traumatic stress syndrome" if there ever is another Nuremburg trial for war crimes.

In Somalia, the Sudan, and Mozambique, the starving masses

are "nutritionally deprived." In the United States, we are loathe to admit that we also have starving masses or homeless people living on our streets, so we reclassify them as "individuals without permanent structural domiciles." During the Reagan-Bush era, "the Great Communicator" told us about a "revenue enhancement" program that looked suspiciously like we were going to have to dip into an "equity recovery program," or a second mortgage, to be able to afford the new taxes. And for reasons probably best left forgotten, when it came to nutritionally-balanced meals, the administration declared that catsup is a vegetable.

Although some students spend their academic years vegetating, even they have a cover up. Students who don't get good grades are "socially or culturally disadvantaged," but there are fewer bad grades now than ever. Grade inflation lets students believe that they've actually learned something when they have a 3.0 average, and that no one will be the wiser when half the class graduates in the top 10 percent. Naturally, if they can find a job, they'll soon learn that an evaluation of "competent" roughly translates as "does the job, but with no great ability."

The 1966 plans for Hampshire College in Amherst, Massachusetts, probably written by a PR person who may have been a journalist but was edited by a committee of pseudo-intellectual cretins, called for the new college's mission to center around (and I'm not making this up) "a social structure [that] should be optimally the consonant patterned expression of culture . . . that higher education is enmeshed in a congeries of social and political change . . . that the humanities offers a surfeit of leeching [and] the exquisite preciosities and pretentiousness of contemporary literary criticism. [Further,] a formal curriculum of academic substance and sequence should not be expected to contain mirabilia which will bring all the educative ends of the college to pass."

From a plethora of psychobabble, often originating within the colleges, have come phrases meant to let us believe all of us aren't "at the margin of mental health." So, we merely need "our space" to avoid "feeling so vulnerable," the result of having had an "emotional disturbance" with our "significant other," and forcing us even more into a state of either being "hyperacidic" or "inner directed" when we should all just "mellow out." To keep our "warm fuzzy" instincts alive, a pharmaceutical chain recently mailed its advertising flyers not to "occupant" or "resident," but to "valued friend." (By the way, you can thank me now for "knowing where you're coming from," "feeling your pain," and for "sharing" that information with you.)

To make sure we all have a "feeling of self-worth," we have

changed job titles, but not responsibiltiies. Mechanics are "service technicians" or "performance maintenance specialists." Trash collectors are "waste management specialists"; movers are "relocation service representatives"; gardeners are "horticultural technicians"; cabbies in the big city are "urban short-run transportation specialists"; and waiters and waitresses are "beverage delivery consultants" or, in the era of being politically-ambiguous, "waitrons." Sort of like what Buck Rogers might have for a pet.

In business, we look to the "bottom line" to "maximize profit" so we have a better "cash flow" and achieve a "high degree of liquidity." Helping business avoid clear language are the legions of PR people, some of them former journalists, who have invented their own secret language to justify their existence. PR textbooks often inform future "strategic planning analysts" or "consulting image enhancement specialists" how to determine behavioral and attitudinal objectives to better "target" the "multitudinous publics," while taking "proactive" stands. To accomplish their mission, the "practitioner" must be able to anticipate, analyze, and interpret; counsel, research, and evaluate; plan, implement, and organize their multifaceted campaigns in order to provide an "intensification of exisiting positive behaviors" and a corresponding "reversal of negative behaviors."

If for some reason, business must "reorganize," "shift responsibilities," or even "go back to square 1," we should realize that they are only in "a state of temporary incapacitation," possibly caused by a "negative economic growth." But, if it continues, then business cites the "bottom line" and the need for "maximizing profits" as justifications for "downsizing" and "restructuring" which, naturally, has caused "a realignment of the work force" and subsequent "involuntary severance." During the past two years, Sears "involuntarily severed" 48,000; General Motors "involuntarily severed" 85,000-100,000; and IBM "involuntarily severed" 125,000. (Somewhere in Bangladesh, thousands of 12-years-olds earning $5-12 a month are making clothes for distribution to Americans who have been involuntarily severed when their companies "restructured" to "maximize profits," and went overseas.) Of course, the only ones who aren't being involuntarily severed by American business are the bosses (pretentiously known as chief executive officers) who caused all the problems to begin with.

JANUARY 1993

Editing the 12 Commandments

Every writer needs an editor, although most writers sometimes wonder why. God and Moses faced the same problem 3,000 years ago. Fortunately, a reporter was present at that meeting on Mt. Sinai. Unfortunately, he could hear only Moses's side of the conversation. . . .

Now, Yahweh, I really have to talk to you about those Commandments. We have an economic crisis at the moment. You see, the price of stone tablets has gone sky high. That means we have to cut back on some of the commandments in order to save money . . . Yes, I know you think 12 commandments are necessary, but it'll cost us more to use three tablets. . . . Yes, I guess you can just make more rocks.

But, Yahweh, there's this other matter. You see, your people have too many things to do to read all those commandments. Right now, they're down in the desert partying and watching a real cool comedy routine from Maimonides Seinfeld. . . . No, I don't think there's going to be a problem with the Jews acting like they were life members of a college fraternity since my brother Aaron is taking care of the people while I'm gone. But, Yahweh, your people don't have much of an attention span, so to hold their interest, we add a couple of charts and pictures, and float them into one killer page design. Of course, it means we'll have to cut some of the text. A few words here. A couple of paragraphs there. . . . I realize that the words are the most important thing, but work with me on this.

For starters, you gave me seven paragraphs just about you. Don't you think that's a bit egotistical? . . . . WO! Hey, Yahweh, could you lighten up with those lightning bolts? I'm just looking out for your own interest. I mean, Yahweh, you say you're the bossman. The Big Kahuna. We all know that. But then you go ahead and say all that stuff about graven images and jealousy and vengeance and swearing. How about we just tighten it up a little. How's this sound?—"I'm in charge and don't you forget it." . . . Well, we can work on it.

Now, the next three paragraphs deal with taking a day off. That's good. Shows the people that you believe in the labor movement. But, *three* paragraphs? Why not just, "Don't overwork yourselves and others so you have time for reflection?"

Now, of the next six, isn't there any you could do without? . . . Oh, I see, it's commandments that the Jews can't do without. Well, what about dumping the "don't kill" commandment? I mean, there's going to be a lot of killing in this world. What about a few exceptions? . . . I see. It's your world and you don't want people killing anything. No exceptions. Well, then, *you* deal with the NRA.

Now, the commandment against that adultery thing could get a bit tricky. After all, a lot of your chosen people chose to go into the entertainment industry. . . . OK, so the commandment stands as you wrote it.

Maybe we could combine the commandments about not stealing and coveting? Seems a bit redundant. . . . Well, that's true. I guess some Philadelphia lawyer could claim that lusting for things isn't the same as stealing them, and some D.A. in the name of justice could tack two charges upon some thief so there's something to work with in a plea bargain. Maybe we dump the false witness thing. . . . I see, you're saying that it's needed because of something known as an O.J., which won't occur for three millennia? Well, if you say so.

I suppose the clause about honoring mothers and fathers stays? . . . Well, sure, if you have this covenant with Hallmark Cards, I'm not going to tell you to renege.

But, Yahweh, we still need to cut you back by half. . . . *Six Snappy Secrets to Success* is a better book title than your *The Daily Dozen*. . . . Yes, I know you have final right of edit. OK, here's the deal. Dump just two of the commandments, let me tighten up the writing on the rest, and I may even be able to get you a film deal; maybe even a big-name star. Add in some special effects, and we'll call it "Firestorms of Desire." . . . O.K., have it your way, but I'm telling you, "The Ten Commandments" just isn't sexy enough.

AUGUST 1995

All in Vein

Some people think columnists spend all morning lying around on the couch, surrounded by beautiful women while contemplating outrageous things to write, then the afternoon writing it up in an opulent office funded by a consortium of lobbyists.

All of that is true, of course, but that still leaves the evening when we're expected to spend long hours sitting with our sources, grilling them incessantly until they admit to having shot the Easter Bunny so we can justify our 800-word pack of lies. Because of extensive research the past few years, I developed a case of varicose veins.

More than 60 million Americans, two-thirds of them women, have varicose veins. For many women, the first time varicose veins pop up is during pregnancy, often because hormonal changes can increase pressure on the veins. But, heredity and a sedentary lifestyle also contribute. Varicose veins can itch, burn, and leave that heavy feeling that makes you believe you're walking on Jupiter. The treatment is everything from a couple of aspirins a day to vein stripping, a painful surgical procedure that, fortunately, has been replaced, except for the larger veins, by sclerotherapy, the relatively painless injection of a solution into the veins which will miraculously disappear.

Armed with my medical analysis, I did what every right-thinking columnist does—I tried to get rich from it. After the Commonwealth maliciously rejected my claim for total disability payments for what was obviously a work-related injury, and told me to come back when I had hemorrhoids, I surrendered to my friendly neighborhood physician. She looked at my legs—"It's what you writers get for sitting on your press passes all day"—and translated "varicose veins" into "chronic venous insufficiency" and threw in "stasis dermatitis" for good measure, thus guaranteeing I could bill the insurance company for a pair of custom-made $90 a pair Lycra-and-nylon "graduated compression stockings." The purpose of the support socks, explained my physician, is to move more blood and oxygen in the direction of my brain. Since some critics have suggested I must have written certain columns while unconscious, I suppose the support socks served more than a medical purpose.

A 20-page manual, which claimed the socks "were easy to put on," gave directions. A lever, fulcrum, and two weightlifters from Ripped-R-Us assisted. Invasions of Third World countries have been planned with less effort than it takes to maneuver the socks past ankles and knees.

Nevertheless, after a couple of hours, I felt light in the legs, convinced I could leap over tall buildings with a single bound. A few leaps later, with the socks easier to put on, I developed a bad case of slippage, a problem far worse than what has happened to the interest rate on my pathetic savings account. In spite of an elastic band at the top, the socks kept falling down. This meant that every hour or so, I'd have to sneak off into a closet somewhere to pull them back up, thus giving rise to numerous wild rumors from fellow journalists, most of whom were either developing or becoming hemorrhoids.

My physical therapist sweetly explained that often the legs become thinner, and the socks don't hold as well because one of their functions is to move fluids out of the legs and into the rest of the body. It explained why my shirts were becoming tighter. A roll-on glue (and how could I *possibly* be making this one up?) was the solution. As for the shirt shrinkage, I'm looking for a body stocking at the moment.

Then there's the washing problem. Normally, I would just throw them into the washer with my usual assortment of towels, pants, ties, jackets and duffel bags. However, the manufacturer warned I'd have to soak the socks in soapy warm water, but not using soap that can be used for lingerie—fortunately, I stopped wearing lingerie months ago—and then hand-wring, pat out, and air dry them.

Thanks to my wife's suggestion, I hung the socks on a hanger from the overhead lamp in the dining room. She said they'd dry faster that way. Now the only problem is that whenever the doorbell rings, I pull muscles leaping up to hide the socks.

After just a few weeks, I can feel my veins shrinking. Soon, I'll be able to call the Eileen Ford modeling agency and see what work they have for smooth-legged journalists.

OCTOBER 1994

For Sale: Justice—O.J. Style

"Get your O.J. T-shirts, mugs, and witnesses here! All shapes, sizes, and colors! Gettem while they're hot!"

Behind a stainless steel vendor cart on Manhattan's Avenue of the Americas was—who else?—Marshbaum who was doing a brisk business.

"Marshbaum!" I shouted, "what are you doing selling witnesses in New York!"

"Because L.A. doesn't allow vendor carts on Sunset."

"That's not the point," I said. "You shouldn't be selling witnesses anywhere."

"Actually," said Marshbaum, "I'm just their agent. I put witnesses together with attorneys and editors, and get 15 percent as my commission."

"We got two forensic pathologists available! Bargain basement price. Only two thousand a day plus expenses. They'll go either way." Two attorneys opened their wallets and walked away with the discounted docs.

"Marshbaum, I don't know what scam you're working, but I think it's illegal."

"Hardly," he said snickering. "It's good old-fashioned American capitalism, and supported by the Constitution."

"The Constitution allows this?" I asked skeptically.

"Sixth Amendment," said Marshbaum smugly. "The right to capitalize on crime."

"Got a psychiatrist here. A little crazy, but does well on the stand. Take her at only three hundred an hour, minimum of four hours." Almost before he finished his spiel, another attorney opened a large briefcase, gave Marshbaum $1,200 in unmarked bills, and bought the psychiatrist.

"You just can't be selling justice like this," I said shocked.

"Who said it's justice?" he said. "I'm selling expertise to the lawyers and information to the news media."

"But only the seedy media believe in checkbook journalism," I said smugly. "The establishment media would never pay a source."

"Got someone here who says he used to deliver diapers to the Simpson home. Only three thousand, and you get 30 minutes with him." A buyer who demanded anonymity put a brown paper bag with low denomination bills on the counter, grabbed the owner of Tidy Didy Diapers, and was last seen darting between traffic on her way back to Times Square.

"The *Times* bought a source?" I asked shocked.

"Can't tell you," said Marshbaum. "But this week alone, I've seen editors from 30 papers. The *Weekly World News* bought an exclusive to a witness who said she could prove aliens soaked up O.J.'s spirit and are using it to cure cancer."

"How long you been doing this?" I asked.

"A month. And if I don't keep hustling, the other agents will grab what's available."

"There's other agents doing this?"

"Of course there's other agents. Why do you think A.C. Cowlings turned down a million bucks? Why did Kato turn down a hundred grand? It wasn't because they're Boy Scouts. They got some smart agent on the West Coast holding out."

"O.K., folks, this one's a little tainted, but it'll wear just as well on TV. She says just before she was arrested for robbery and assault with a deadly weapon, she saw O.J. buy rubber gloves from the store she hit." Three editors tore each other apart trying to get the highest bid. When they were through, an ambulance carried them away, but Marshbaum had $15,000—15 percent of it his commission—for the thief's guarantee that she'd tell the truth.

"That's terrible!" I said, still shocked by the display before me.

"You think *that's* bad," said Marshbaum, "two days ago, I had to have the cops clear the sidewalk when the staffs from "Inside Edition" and "Hard Copy" began fighting over a witness who says not only did he sell O.J. the white Bronco, but heard him say it's a killer machine."

"Jurors! Live jurors! Get the exclusive story of what went on in deliberations! Only fifty thou apiece! Available only to the news media!"

"That's ridiculous," I said. "They didn't even pick the jury pool yet. How can you be selling jurors?"

"They choose jurors from the voting lists," said Marshbaum. "In the past week, I've signed exclusive letters of intent from every voter in L.A. County. Hundreds have been registering just in the past week alone. It's only a matter of time until the lawyers choose who they want on the panel."

"Is there *anyone* you *don't* represent?" I asked.

"Yeah. I don't represent lawyers. Even agents have ethics."

AUGUST 1994

Dysfunctional Thoughts

Is anyone else tired of hearing people claim they were from a "dysfunctional family" in order to justify their life's problems? Have a bad-hair day? It's all because of the gene pool in your dysfunctional family. Greedy and self-centered? You're from a dysfunctional family. Nasty to your neighbors and coworkers? Dysfunctional family. Abuse your wife or children? Yep, Dysfunctional Family Syndrome. Rob a bank? Kill a cop? Just chalk it up to having come from a dysfunctional family.

And, speaking of dysfunctional families, does anyone know what the Queen of England does? So far, thanks to television, for the past four decades we've learned she dresses well, and can ride a horse and wave at the same time. But, what *else* does she do? Elizabeth II seems to be a nice lady, but she could easily be the poster queen for what happens when generations of royal families intermarry.

Is anyone else tired of seeing the news media salivate all over their VDTs whenever anyone with a royal moniker does anything more than just wake up in the morning? Are more people than me suspicious of journalistic quality when our daily newspapers put royal sex scandals on the front page and *People* magazine devotes at least a full page every week to the "Royal Watch"? I wonder how many stories were spiked by the editors in order for them to splash the news that Fergie and Di sunbathed topless, that there was a major scandal in both Di's and Charles's affairs of state, and that some rich American consort was caught sucking toes? Schlock romance novelist Barbara Cartland and the editors of the *National Enquirer* not our mainstream media should be the ones worrying about which royal retard is fooling around with which commoner.

Prince Charles who, if he weren't heir-designate to the throne would probably be just another tweed-suited geek with a bagload of money, voluntarily pays $1.25 million a year in taxes on income of about $5 million. However, the Queen earns $50-60 million a year, and the State pays for all of her living expenses, including five official residences, the planes, helicopters, and Britannia, a 412-foot yacht that carries a crew about the size of the population of Cleveland. Beginning last year, the Queen voluntarily agreed to pay taxes on her income. Normally, that would be about $20 million, but the Magna Charta, or some such unenforceable document, allows her and the government to figure out an appropriate queen tax. The "appropriate" tax is estimated at $3-5 million a year.

At least there is a way for England to rid itself of its debt. In America, our debt is piling up at the rate of $13,000 a *second*. Now, to someone making $50 million a year, $13,000 may not be much, but $13,000 a year is the amount that distinguishes a family of four as being below the poverty level. It'd be boorish for us to ask the Queen for a few bucks, especially since the royals are now quite happy about not having to defend the actions of its former colonies.

However, I have a couple of ideas about how we can rid us of debt.

Illusionist David Copperfield has made a railroad car, an airplane, and the Statue of Liberty disappear. Maybe it would be worth it to pay him a few million to work on the national debt.

If magic doesn't work, and how Congress appropriates its pork can only be described generously as magic, maybe we could package Di and Charles and Fergie and Andrew into an afternoon soap, add a couple of hunks and hunkettes, then sell them all to some American network. My first thought is to drop "Days of All My Royal Romances" onto PBS since the national beg-a-thon network goes into orgasmic veneration over anything British. But, alas, we'd want to make money off the deal. So, we must pitch the other networks.

With the cost to produce a one-hour TV show averaging $1.5 million an episode—with *Roseanne* topping the list at $3 million—and figuring 22 shows a year, that's at least a $33 million outlay. But, being producers, we do what every other producer does—we

skim off the profits, and take our rightful $10 million a year. At that rate, and assuming we can keep the debt from increasing more than $13,000 a second, we should be debt-free by, oh, about the time the TV networks get an original idea.
OCTOBER 1994

Playing the University Grantsmanship Game

You're probably not going to believe this, but a scientific foundation once awarded a $5,000 research grant to a behavioral scientist to study the television viewing habits of cats. The reasoning behind the grant, said the Society's spokesman with not even a chuckle, is that animals, especially cats, react to voice and picture patterns on the television screen, although they don't comprehend what is taking place. The spokesman even went so far as to suggest that cats prefer to watch only good television programs and shun the poor ones.

While the officers of the Society aren't going to have many problems—all they have to do is wait patiently for the report and try to act serious about the whole thing—the poor media researcher, eager for tenure and promotion, is likely to be burdened with uncountable problems.

The researcher, a portable laptop computer loaded with statistical programs strapped around his shoulder, timidly approaches the cat.

"Hey, Man, what are you doing in my sandbox?"

"Oops. Sorry. Ahem, well, I'm your average ordinary university media behavioral research scientist, and I've come to interview you about what television programs you watch."

"You feeling all right? Humidity not too high?"

"No, honest now, Cat, I've really come to find out what you like on TV. I bet you watch cartoons. Felix the Cat, Krazy Kat, and Fritz the Cat!" The cat yawned, so the researcher tried another direction. "You scout the opposition! Lassie? Rin-Tin-Tin?"

By now, the cat was convinced the researcher was a few numbers short of a full equation, but the researcher was persistent. "How about comedies? Everyone loves SitComs!"

"Cats aren't everyone."

"It's adventure you like!" beamed the professor. "Maybe a jungle show with lions?"

"Only if the diet is media researchers."

"Drama? Science Fiction? *Star Trek* is all over the tube. Surely, you've seen at least one episode!"

"You're spaced out," said the cat unsheathing his claws.

"I know!" shouted the professor, "you're a socially-aware cat. You watch *60 Minutes*, *20/20*, and *Nightline*!"

"You're loony," the cat replied, waiting for a chance to dial 9-1-1 to ask the Media Researcher Unit at the state hospital to make a house call. "I don't watch TV."

"But you've just *got* to watch television. It's the American way!"

"Look, Man, like I've been trying to tell you, none of us cats watch television!"

"*None* of you?" came the startled reply of the media researcher, his Ph.D. dripping with an unused grant. "Then what's the TV doing by the couch?"

"Oh, *that*," grinned the cat, "I like to sharpen my nails; that's the only good piece of wood I can find. It's about all that it's good for."

"Look, Cat," the researcher pleaded, "I've got this really neat grant—$5,000 from the Society—and it's to find out what cats watch on television. And besides, I can get a publication credit and—"

The researcher didn't even finish his sentence before the cat's bright green eyes flickered. "You mean some nuts gave an even bigger nut five thousand bucks to interview cats to find out what we watch on the scratching post?" And then, the cat began to think. "Let's see, now. He gets $5,000 to interview cats. And, after all, I am the most important part of the study." He decided it was only right for him to help. "Look, Prof, like I said, none of us cats watch television. But, I'll tell you what I'll do. You keep asking questions. I keep denying them. You're bound to miss at least one or two shows. If I can't deny I don't watch what you don't ask me, who'll be the wiser?"

The media researcher didn't have any idea what the cat said, but it sounded like a plan to him and just as understandable as what media researchers write. So, once again this media researcher was happy. His career wasn't ruined. There would be a publication credit. There would be tenure and promotion.

In an exclusive section of some American city, there walks a very, very happy professor . . . and a cat with a $5,000 bankroll.
AUGUST 1993

The $6 Million Journalist

Journalists are confined by having to report reality. But, often we wish we could report stories that don't exist, but *should*—such as . . .

• Pop vamp Madonna today announced she plans to wear clothes on her next world tour. The Material Girl's "Justify My Blonde Ambition Garment Tour" begins April 15 at L.A.'s Salvation Army Value Store and concludes two months later at London's Carnaby Street.

Representatives of the mass media decry Madonna's plans to be clothed as "another cheap publicity gimmick," and vow to picket her shows.

• After two days of heated discussion, members of the Association of American Publishers voted to "significantly reduce" the number of ghost-written movie star books and to increase publication of books that focus upon important social issues. The Association has also voted to limit publication of the number of purple-prosed romance books, mostly inaccurate weight-loss and personality-make-over books, and all books with the name O.J. anywhere in the text.

• In the spirit of the book publishers, NBC surprised its affiliates with its early morning announcement that it renewed "I'll Fly Away," the critically-acclaimed TV series that focuses upon the nation's civil rights struggle during the early 1960s. The series finished 86th of 93 prime time series.

"While it's true that only six million households a week saw the series when it originally aired," said network VP Seymour Schlock, "we have a responsibility to enlighten the masses to help them better understand their nation's heritage."

• Central High School says that it has reached "an all-time high" in graduating students who are literate. "We hit 75 percent this year," said principal Ed Cashion. Unfortunately, one-third of the faculty are being required to attend summer school to take remedial courses in the 3 R's since they spent more time in college taking education courses in *how* to teach than in learning *what* to teach.

• KFAD-TV, impressed by the work done at Central High, fired news anchorbabe Susie Sweetwater. "Although she was gorgeous, smiled better than any other person on the air, and didn't do too bad with one- and two-syllable words," said news director Homer Simpson, "Susie just wasn't a journalist, and we have exceedingly high standards for reporters in the local market."

• The *Wattabago Tribune* signed award-winning investigative reporter David Bergman to a three-year $6.4 million contract. Bergman, who had been the clean-up hitter with the Philadelphia *Daily News* the past four years, was granted free agency status in January.

"We believe in the American philosophy of paying employees what they're worth," said *Tribune* publisher Arthur M. Greeley.

In a related story, Philadelphia Phillies pitcher Harry Horsehide became the highest paid person in sports when he signed a three-year contract for $76,000 a year.

• And, finally, Avarice K. Toadstool, president of Amalgamated Conglomerate Media Industries, having finally bought up all book publishing, newspaper, and magazine companies in the country, says he finally understands why his employees need better working conditions. From his company headquarters in Bermuda, Toadstool says his company won't fight the effort by his 25,000 Third World employees to unionize. "They deserve representation," says the CEO, "and I encourage my loyal employees to avail themselves of the better benefits, working conditions, and grievance procedures the union can provide."

MARCH 1993

Stealing the First Amendment

"Got a match?"

I didn't know where he came from, but there he was, right behind me—as usual. "You know I don't smoke," I told Marshbaum. "Come to think of it, you don't either. What's up?"

"Not much. Planning to roast some marshmallows and hotdogs. Burn some books."

"Marshbaum," I shouted. "You can't burn books."

"Sure I can. All I need is a match. See, first you—"

"Burning books is against everything this country stands for."

"Not when the books are evil."

"Marshbaum! Didn't you ever read *anything* Jefferson wrote? Our country was founded upon the principle that all views must be heard."

"Sure, and my view is that we're going to burn some books to keep them from causing any more trouble." I could have given Marshbaum a 10-minute lecture about John Milton's arguments that those who seek to destroy books destroy reason itself, and that mankind is best served when there is a "free and open encounter" of all ideas. Or, maybe, a few words from philosopher John Stuart Mill who stated, "We can never be sure that the opinion we are endeavoring to stifle is a false opinion, and if we were sure, stifling it would be an evil still." Maybe a little bit of wisdom from supreme court justice Oliver Wendell Holmes who told us that democracy is best served in a "marketplace of ideas." But, I knew Marshbaum wasn't in a mood to hear philosophy. So, all I said was a sarcastic, "I assume you plan to burn everything you think is evil."

"Just romances. Historical. Contemporary. Anything with a female byline and a cover of a woman with lust in her eyes and a torn dress on her bod."

"Why Romances?" I asked.

"Because Romances lead to crime."

"Most Romance novels may be a crime against good writing," I said, "but that's still no reason to burn them."

"It's a case of numbers," said Marshbaum. "The major publishers put out more than 120 Romances a month, three times more than 14 years ago. And, the average Romance reader spends about $1,200 a year on the books." He smugly told me he got that information from the Association of American Publishers.

"Wasting money on syrupy nonsense still isn't enough of a reason to burn books," I said.

"Get away from your computer and see what's been happening to America," said Marshbaum. "In the past 14 years, the crime rate has also tripled. It's a direct relationship. Where do you think those bored housewives get all that money to buy Romances? You can't read just one Romance. Once you're hooked, you need 10, 20. You become addicted. You need more and more until one day your whole life is nothing but a long print-enhanced haze of bodice-rippers and sap. Eliminate Romances, and you'll walk safer in Central Park at night."

"That's the most convoluted piece of logic I've ever heard!"

"The *Washington Post* says it's OK to burn books," said Marshbaum.

"I doubt the *Post* believes in burning books," I said eruditely. I was eruditely wrong.

"Well, maybe not books, but newspapers. Remember all the thefts of newspapers on college campuses?" I remembered them vividly. At more than three dozen campuses a year, according to the Student Press Law Center, students who disagree with something in the student newspapers steal them from distribution racks, and throw them into dumpsters or burn them. The culprits were often persons who thought of themselves as liberals, but whose actions certainly suggest they spent more time in parties than in classes that discussed the founding principles of the nation.

"So, how does all this tie into the *Post*?" I asked.

Marshbaum pulled an editorial from his pocket. "Read this!" he commanded. According to the *Post* editorial, dated January 22, 1994, "scooping up copies of publications—whether to send a message or to protest one—may in itself be a form of free speech." It argued that proposed legislation in Maryland to ban the theft of newspapers is "neither a good idea [nor] a worthy pursuit" by the legislature. It was a shocking philosophy from a newspaper that screeches a national emergency almost anytime a governmental agency doesn't yield all its secrets.

"Obviously someone kidnapped all the reporters and editors," I said.

"Obviously the *Post* has divine guidance to determine what is truth," said Marshbaum. "Maybe *they* have a match I can borrow."

Against the Post's *advice, the Maryland legislature became the first state in the nation to declare it illegal to steal newspapers with the intent to prevent others from reading them.*
JUNE 1994

Creating Historical Fiction

A few years ago, I tried to tell my younger son that some passages in his 9th grade history book were incorrect. He refused to believe me. Afterall, there it was, written in print and paraphrased in lecture by his teacher. Even if he believed me, he knew he'd lose points on the next test if he didn't answer the way his teacher and textbook expected.

Even in the most meticulously-researched and written text, errors of fact are inevitable. However, because many public school social science texts are little more than well-designed cut-and-paste jobs from other texts, and often put together by committee, the errors in one are often compounded in the next. Nevertheless, errors of *commission*—a wrong date, the wrong name, an interpretation not based upon facts—can usually be fixed either by an addendum, circular to teachers, or in the next edition.

Errors of *omission*, however, are more serious because they can't be verified as easily, and may have been caused by author, editor, or publisher bias, ethnocentrism, or just plain lack of knowledge of an area. Until recently, textbooks routinely omitted significant references to contributions of persons other than White males to the development of the country. Also missing from public school texts was almost anything that cast a negative image of American values, the school boards possibly reasoning that those "darling impressionable minds" couldn't handle all that conflict. Thus, it isn't surprising to learn that textbooks fail to report that Revolutionary War patriots, even long after the Constitution was adopted, routinely violated the principles of the libertarian revolution by significant abuses of civil liberties and of continued violations of freedom of speech and of the press . . . or that both the Mexican-American and Vietnam wars had massive protests of American foreign policy . . . or that the government systematically and maliciously violated due process and civil rights during the McCarthyism witch hunts of the 1950s.

The reality is that the majority cultures write the histories, and their texts often reflect their biases and political agenda. During the latter twentieth century, Japanese texts overlooked the slaughter of thousands of Chinese civilians; Soviet texts failed to mention America's massive economic and humanitarian assistance to that country; and the texts of all countries reported little about the Holocaust. Publishers in America, trying to reap the widest possible financial benefit by not offending anyone, often force authors to overlook significant social trends.

To establish standards for the study of history in the public schools and to correct some of the nation's textbook wrongs, the National Endowment for the Humanities, under Congressional mandate, gave $1.75 million to UCLA's National Center for History in the Schools to bring together a wide range of academics to study the problems and to recommend a model text that would present history as it was, rather than what we hoped it was. The concept was good; the execution was abysmal.

The Center rightly determined that texts were "sugar-coating" and distorting American history, that there was an overemphasis upon a recitation of facts and in recounting the deeds of a few people, mostly white males, but far too little discussion about major trends and social issues that defined the American republic.

However, in its recommendations, the Center did exactly what it condemned. With a political and social ideological agenda, the new standards, presented in a 271-page document at the end of 1994, discounted the Western European influence in the formation of the United States, and presented a distorted overview that the formation of the country was a convergence of Islamic, African, and European influence. It claims the nation's continuing history is little more than struggle, conflict, and the abuse of the rights of people. It barely discusses the historic role of a free press and of free speech, mentions the Gettysburg Address only briefly, and relegates the complexities of the "cold war" as merely a "quarrel" among imperialistic nations.

The Committee's proposed guidelines, although rightly adding many civil rights leaders, left out Eli Whitney, Thomas Edison, and the Wright Brothers among many other scientists; it overlooked Daniel Webster and other major diplomats and politicians; and it gave few lines to innumerable creative artists. However, the emperor of an ancient African civilization is praised, as are numerous individuals, often female or of minority cultures, who were merely footnotes in America's 300-year history.

In historiography, as in journalism, the better writers not only research the facts, but also analyze and interpret them to help the people better understand the critical issues that affect them. It is not acceptable for the writers of public school social science and history texts to distort the reporting of history by creating fiction.

MAY 1995

The Press Meets the Afo-a-Kom

The return of a ritualized and functional royal throne figure of African heritage from a New York City art dealer to the Kom people of Cameroon was widely but poorly reported by the world's media in late 1973. Not much has changed since then.

The Kom story was part of the budgets of virtually all American and several overseas wire services. News and feature articles were soon appearing in most of the world's media, including *The New York Times*, *Washington Post*, *National Geographic*, *Die Bunde Illustrierte*, *Ebony*, and *Esquire*, as well as the three major American television networks. The Afo-A-Kom even became the answers to questions on a number of television quiz shows, including "Jeopardy." The story was hot, and the media knew it.

Depending upon what was read at the time, the Afo-A-Kom, a hundred-year-old statue, was either a god . . . a phallic symbol . . . an icon . . . or a symbol of tradition, unity, and harmony. It was either "as sacred as" two other Kom statues . . . or the most sacred of all. It was either a national shrine of the Cameroon . . . or a part of the Kom people. It was 5- or 5-1/2 feet tall, or 62-, 62-1/2 -, or 64-inches tall. It was made of ebony . . . wood . . . iroko. It was covered either by red beads . . . or reddish-brown and blue beads . . . or brown and blue beads. The beads were coral . . . plastic . . . semi-precious stones.

It was stolen from the people of Kom in 1966 . . . or 1967. It was taken from a storage hut that was loosely guarded . . . or not guarded at all. The original thief—there were many—was the son . . . or the nephew . . . of Fon (King) Law-Aw . . . or Al-Aw . . . who died shortly thereafter . . . or who died before the theft. The thief was ostracized . . . punished . . . not convicted. The Fon himself was powerful . . . or not-so-powerful . . . or had no power.

Aaron Furman, a New York City art dealer who had either an unblemished reputation . . . or was a scoundrel. . . purchased the Afo-A-Kom in Cameroon . . . or Paris . . . or somewhere else. In 1967 . . . or 1968 . . . It was brought to the United States, with Furman apparently believing that it was legitimately purchased . . . or knowing it was stolen. It was valued, by American standards, at $30,000 . . . $51,000 . . . $60,000 . . . $65,000 . . . $80,000. Furman eventually sold the statue for his expenses" . . . or "a substantial sum" . . . or $25,000.

According to the media, Kom is a series of villages . . . a tribal enclave . . . Shangri-La. But, no matter what it is, Kom is located, according to the popular media, in West Cameroon . . . or East Cameroon; in the Federal Republic of Cameroon . . . or the United Republic of Cameroon. As for the people of Kom, that remote Cameroonian kingdom, they number 3,000, 35,000, 40,000, 81,000 or 92,000. No one was really certain. The Kom, stated the media, are also primitives . . . savages . . . or trapped by modern civilization.

The first story about the Afo-a-Kom broke in the October 25, 1973, issue of *The New York Times*, after one of its correspondents had learned from two Peace Corps workers of the Afo-a-Kom's theft. Within hours, the other national media jumped onto the bandwagon and became overzealous, rewriting and compounding errors. Even the *Times*, which published 15 stories (about 700 column inches) over a two-month period, committed numerous errors of fact and interpretation.

When the media weren't bungling the facts, they were locked in a trap of ethnocentrism. The majority of articles that came from the American media showed a basic lack of understanding of cross-cultural communication, as well as a certain lack of understanding of the people whose beliefs, world views, and languages and cultures are not those of certain Americans.

Not untypical of the views were those of a writer for *Esquire* who apparently overdosed on *Tarzan* movies. Armed with experiences of less than a half-day in Kom, a place she never knew existed prior to 1973, the writer claimed she was going to write a "definitive" study of the Kom people." She claimed she had been "to the edge of the earth." She admitted, reflecting the views of many reporters, "I wanted to go to Kom. I wanted to see the Fon's fifty wives and half-naked slaves and all the pomp and panoply of great African kingdoms, where the people would receive us white *bwanas*, bowing low or maybe even casting themselves to the ground and throwing dust on their heads. I've heard of things like that. Also I wanted to see the Fon ride out on his legendary pure white horse to greet us, followed by ten servants carrying calabashes of palm wine and his umbrella and his chair." She worried whether the salad was safe to eat, called the flora of Cameroon "strange plumy vegetation," and called Kom's history "bloody murder, war, witchcraft, fire, famine, and flood," without recognizing that the history of the Kom is no different than the history of any people, including the Americans.

The *Times*, with numerous reporters parroting its reporting, inaccurately claimed, "the women of Kom are completely subordinate to their men. They are expected to work hard, bear many children and not complain. The women do all the farming, cooking, cleaning, child-raising and most of the market selling. . . . Many women say they like polygamy because it lessens the very heavy work burden." The Kom, however, are a matrilineal society, and there are clear-cut distinctions in Kom society between sexual, biological, and sociological wives.

In contrast, some of the media, especially BBC and the *National Geographic*, were relatively accurate, taking the time to verify facts and evaluate the sources of information.

Why the major media gave the story such heavy play probably is a reflection of Americans' fascination with "exotic" lands, crime and conflict, and what we thought was a work of art that carried a price tag. Johnson Ndimbie, cultural affairs officer at the Cameroon Embassy to the United States and Canada, believes the story received the coverage it did because not only were the media trying to find stories about Africa, but that the media are "more concerned, or more capable, of covering the bizarre and sensational. They seemed to be more interested in coups and conflicts because it makes good reading."

In contrast to the exploitation of the story by the larger media, the smaller papers usually spiked the wire service copy. Most newspapers argued, "Well, you know that there's a newsprint shortage and we really don't have the room. Also, no one is interested in things happening in Africa. It's just too far away."

And that's the problem, but it isn't solely one caused by the media; they merely mirror society.

JANUARY 1975

The Case for Sensationalism

Within the past few weeks, almost every American newspaper, news magazine, radio and TV station has hurled stories at the public about Streptococcus-A, a flesh-eating killer infection which can cause death within a couple of days. It's a story the media love, just as they love stories about sex, violence, and the marital affairs of film stars.

The critics say the story has been overplayed, accurately pointing out that the incidence of the killer infection, which acts as a toxin to the body's muscles, is no greater this year than any other year. They say the media are pandering to the public's worst fears in the hope of gaining higher readership, and that exaggerated and sensationalist reporting are causing a panic that is resulting in physicians and hospitals being jammed with people who may have no more a problem than a mosquito bite.

All of the criticism is probably true. But it was an historical precedent that helped set the base for the public's right to know, even if the media do occasionally enter the "grey area" of reporting.

It was 1834, and the upstart *New York Sun* was causing trouble for city politicians and fellow journalists. While other newspapers charged six cents an issue, made their readers buy a year's subscription, and their advertisers a year's contract, the *Sun* sold each issue for just a penny and allowed advertisers to purchase their ads whenever they wished. The other papers scorned such outrageous business practices.

The more established papers, targeting the city's business class and the elite as their readers, accepted bribes and favors from politicians and the public to keep their names out of the paper or to print "dirt" on some opponent. Naturally, they never reported that judges, police, and politicians also routinely took bribes, some from the newspapers themselves. But, *Sun* publisher Benjamin Day and newly-hired editor George Wisner, with a slate of police and courts news, looked to the city's lower- and middle-classes for their support; the *Sun* argued it would print all the news or none. In response, the other journalists brazenly told Day and Wisner they were upsetting accepted practices that didn't seem to hurt anyone, yet helped fellow reporters and editors make a respectable wage.

While other newspapers kept their reporters in Manhattan, Wisner hired "stringers," community-based people in different parts of the city to report news when it happened, earning the wrath of journalists who, apparently with straight faces, called such practices a blemish upon the profession.

Soon, the *Sun* attacked filth in hospitals and prisons, lashed out against slavery, opposed monopolies, and argued that an artificial increase in the price of flour was nothing more than greed. The establishment press retaliated by demanding a grand jury investigation on charges of inciting trouble for business. That demand, as well as numerous suits for libel, were usually thrown out of court.

Less than a year after becoming editor, Wisner learned that a New York City resident had died of cholera, an extremely debilitating illness that attacks quickly and, at that time, had a relatively high mortality rate. The cause could have been in the city's water supply or in tainted food sold in grocery stores or in restaurants. Wisner knew if there was one case, there had to be others. But, the public health department said there was no problem with cholera, certainly no epidemic.

Distrusting bureaucrats as much as he did the competing newspapers, Wisner asked his stringers to do a little bit of "leg work." When they reported there were other cases, Wisner went to the health department, which again denied any problem. After some badgering, Wisner got the health officials to admit there "may" have been a problem. But they said if the *Sun* published the story, there would be a panic that would lead the public to needlessly tie up doctors and hospitals. Besides, said the health department, the other, more "responsible," newspapers, knew about the potential epidemic, and had kept quiet because it was, said the health officials, "in the public's best interest."

The public's best interest is to know all the facts, said Wisner who published the story and suggested the health department was negligent in detecting the disease in the first place. The other media attacked him for being irresponsible. But that aggressive "irresponsibility" led to the people knowing what was happening in their society and to their lives.

Slightly more than a decade ago, the media were asleep when the first reports were made about an infection that had no known cure, had the potential for epidemic proportions, but was confined to a small group sociologists and the media called "deviant," but most of the public called "perverted." The infection became known as AIDS.

Had the media understood George Wisner's philosophy of journalism, the people would probably have known about AIDS a half-decade earlier, and may have been able to take evasive measures that would have saved thousands of lives. With today's Strep-A story, perhaps there was media overkill. But, mankind is largely rational, and it's better to have all the facts than to live in ignorance at the mercy of things unknown. If the media hype caused one person to recognize the symptoms and to seek treatment, the coverage was worth it. If thousands of people who didn't have Strep-A were scared enough of suspected symptoms, and went to see their physicians who may have diagnosed and treated other less serious illnesses, the hype was also worth the ink. Although the media can rightly be accused of pandering to the public's fears, maybe it's permissible for there to be some "sensationalist" reporting now and then.

JUNE 1994

Make Mine Media Rare

Whenever people get upset with the media, they often demand that reporters be licensed. When challenged that the First Amendment and America's libertarian philosophy forbid governmental bodies to determine who can and can't report the truth as they see it, these critics become upset. If reporters wish to be treated as professionals and be taken seriously, these critics whine, then they should meet certain educational, ethical, and professional standards, like lawyers, teachers, and beauty shop operators, all of whom must be licensed.

But, in our emerging era of Newtonian democracy, the Constitution may get tattered a bit. And so it was that Sam Frattlebaum, a journalist for 25 years, was the first to be brought before the Court of Ultimate Decisions of the State Board of Journalistic Licensing.

The chair of the three-member court looked down upon Frattlebaum, and solemnly asked him how he pled to the charges of malfeasance, gross negligence, and reckless disregard of the truth.

"Not guilty," came the whimpered response.

"How *dare* you!" thundered the chair. "We wouldn't bring you here unless we were sure you were guilty! Now, stop wasting this court's time, and admit your guilt."

Once again, a frazzled Frattlebaum pleaded not guilty, forcing the chair to present the case.

"By 8 a.m. the day after the murders of Nicole Brown Simpson and that other guy whose name none of us remember 3,000 reporters were all over L.A. Where were you?"

"I was in Rwanda covering the massacre of the Tutsis," said Frattlebaum.

"No excuse," snapped the chair. "A half-million dead people in Africa aren't as important as one American football player. Where were you during the preliminary hearings?"

"I had to return to Bosnia to report about a new wave of Serbian ethnic torture."

"Don't you think O.J. is tortured by these charges? During six months of preliminary motions, you again deserted your country."

"Your Honor, I was in Japan to report about the earthquake that killed more than 5,000."

"Did you find any of Judge Ito's relatives? Maybe dig up some sex scandal?" Frattlebaum hung his head. "The record also shows you weren't even in L.A. when the most important trial of the millennium began."

"No, sir, I was in West Virginia to interview some hill people for a series about hunger in America."

"Don't you think O.J. gets hungry? What about the prosecutor's favorite foods? What's the judge's eating habits? What's the Defense team's favorite restaurant? Since ice cream is such an important part of the trial, didn't you want to find out its flavor? Was it ice milk or frozen yogurt? How many ounces? What time did it melt? Certainly, you could have interviewed Ben and Jerry for a feature profile."

"I didn't think that—"

"It's obvious you didn't think," snapped the chair, continuing his inquisition. "You also didn't think about bribing lab techs to reveal the DNA results or to badger some cop down in Homicide." He checked his notes, then actually smiled. "Ah, I see you did try to meet your responsibilities, and finally made it to Los Angeles."

"Yes, sir, I went there to report about the people left homeless by the floods."

The judge sighed deeply. "I suppose you didn't have the insight to realize that a flood could damage orange trees, lead to higher prices for orange juice, and affect the public's perception of O.J."

"No, Your Honor, but I did get a story about gang killings. Do you know that last week in L.A., 65 people were murdered, 12 of them children?"

"I don't care if the whole voter registration list was snuffed. None of them were TV sports commentators who run through airports huckstering rent-a-cars."

"I did try to get to the trial one day," said a sweating Frattlebaum, but I tripped over some TV cables and had to go to the ER."

"Yeah yeah," said the chair throwing newspapers in the air. "While in the hospital, you didn't even steal O.J.'s medical file. When they let you out, did you at least try to interview a juror? Sneak into the O.J. compound for exclusive photos? Plant a hidden microphone in the judge's chambers? Find out if the judge wears briefs or boxers. The public has a right to know!"

"I'm sorry," said Frattlebaum, "but I had to leave to cover the story of a family who were being evicted after they exposed the worst case of corruption in Housing Authority history."

The chair sadly looked at Frattlebaum. "Your failure to report the important stories is killing our credibility. By your own words, you haven't lived up to the standards set by your peers." And with that, he slammed down his gavel, and suspended Frattlebaum's license for one year. "Maybe in that time," said the chair, "you might develop a sense of what's expected of our profession."

FEBRUARY 1995

An Obscene Story

Throughout the country, people are creating barriers to what they believe is obscene, trying to ban everything from rap records and music videos to works of literature, newspaper columns, and museum art. Even the courts have become involved, changing definitions as quickly as chorus girls change costumes. Naturally, no one has declared the multi-billion dollar savings and loan scandal, the selling of arms to Iraq and Iran, or what's happening in Bosnia to be obscene.

Nevertheless, the recent upsurge in obscenity legislation and prosecution had to begin somewhere. By the time it ended, Sidney Thornacre, a mild-mannered stock clerk from Driven Snow, Pennsylvania, was elected President of the United States without opposition and on a 1-0 vote. It was 1-0 in the popular vote and 1-0 in the electoral college, the first time that anyone was elected president unanimously. How President Thornacre was elected shall remain an inspiration to all people throughout the world.

It all began when the Driven Snow City Council decided that

there had to be laws against obscenity. But the Supreme Court had long ago determined that for laws to be fair, they had to be spelled out—in painstakingly exact detail. According to the Supreme Court, it wasn't good enough just to say that obscenity was bad, the laws had to specify just what was bad.

So, the City Council read the appropriate books and magazines, watched the appropriate television shows and movies, looked into the Internet, and spelled out a series of ordinances so explicit they even made Hugh Hefner blush.

Naturally, the laws themselves were obscene and all members of the city council were sent to jail for having written them. After all, whoever is a part of obscenity must be punished. And that's how the DA and his entire staff, who had to first read the ordinances to determine what it was they were prosecuting, were prosecuted and sent to jail.

One after another, the prosecutors became the prosecuted. Although many recognized what would ultimately happen, they had a duty to perform. After prosecuting the guilty for reading the obscenity statutes, they willingly accepted their own prosecution. America's sense of family values had to be preserved. Soon, the entire population of Driven Snow were in jail, victims of rampant obscenity. From Driven Snow, it spread throughout the rest of the country.

And then it happened—something everyone feared but no one expected. The President of the United States, hoping to stop the problem, read the ordinance, admitted guilt, then resigned, leaving the country without a President or government—the Congress had been among the first ones to read the ordinance.

And that's how Sidney Thornacre, the only person in the country who hadn't read the ordinance, applied for the Presidency. On that fateful day, he alone went to his precinct polling place—where he was precinct captain, judge, poll watcher, and election counter—and cast his lone vote. Then, he made that vote official by going to the nation's capital; as the country's only member of the electoral college, he cast a unanimous vote for himself for president.

On a wall of the Oval Office, Sidney Thornacre—President of the United States, Commander-in-Chief of the Armed Forces, secretary and staff of all the cabinet departments, Chief Justice of the United States, lone member of Congress and House janitor—has a framed copy of the anti-obscenity ordinance—with appropriate sections blacked out.

NOVEMBER 1992

Of Matrons and Movies

Three middle-aged women, inhibited but longing to find out what the new morality was all about, snuck out of their homes early one afternoon and went to a movie.

The movie was shot in Sweden, but the subtitles were in English. Not many in the audience, other than the three ladies and a small group of cinema students, really cared what the subtitles said.

"They didn't tell us it was a foreign film," protested the first lady.

"I think they're talking Danish. You know how the Danes are," explained the second lady.

The third lady just giggled.

The movie was about a young girl who believed she was frigid. A rather frightening experience for a 14-year-old. So, she went to a psychologist for help. Soon, the psychologist and the girl were in bed together.

"Horrible!" cried out the first woman.

"Disgusting!" protested the second woman.

The third woman just giggled.

The girl, "searching to find herself," traveled to Germany where she met a young student. The two of them did a lot of studying together.

"It's the Supreme Court and its radical rules!" declared the first woman.

"It should have been banned!" declared the second woman.

The third woman didn't declare anything. She was too busy watching the action on the screen.

However, this was a well-filmed movie with good color and adequate acting. By Supreme Court standards, it was "socially redeeming."

In Italy, the young girl met a fisherman, and in Spain, she met a poet and a prostitute.

"I won't let *my* family see this!" huffed the first woman.

"Totally shameful," huffed the second woman.

The third woman huffed nothing. It's hard to huff when you're giggling.

After Spain came France. By now, the young girl had an armful of books she had collected throughout the Continent. She had collected the works of Heine, the *Story of Raphael*," and a libretto from *Faust*.

"See! She's adding to her collection of dirty books," snickered the first woman.

In France, there was a vulgar show in a respectable-looking nightclub. The young girl became ill when two women wrestlers began rolling around in the mud. But when the young girl left, she accidentally ran into the city's most notorious district. Frightened, she ran until she bumped into a well-dressed respectable-looking well-mannered gentleman who offered to take her away from the sordid life. They left in his Rolls-Royce.

"He'll save her!" joyfully proclaimed the first woman.

"He's so respectable. Maybe that's the message," happily proclaimed the second woman.

The third woman was having trouble controlling her laughter.

The man was a sado-masochist. Looks are deceiving.

"Shocking!" cried out the first woman.

"Disgusting!" cried out the second woman.

The third woman just—yeah—giggled, but it may have had nothing to do with the film and everything to do with her companions.

But the first woman who said "Shocking!" and the second woman who said "Disgusting!" and the third woman who giggled, stayed until the end of the film.

Later, the woman who giggled her way through the girl's European adventures led a citizen's protest at a local video store.

JANUARY 1994

Wonderings of an Idle Mind

A lot passes through a columnist's mind while waiting for those great thoughts to lead to a weekly column.

For instance, we all know that Superman is a journalist—and, of course, most journalists are super men and super women. But, does anyone know how Superman gets all that time off to pursue truth, justice, and the American way? Has anyone ever actually seen Superman attend a city council meeting or even a PTA meeting? When was the last time you actually saw him writing a story against a deadline?

Bruce Wayne, of course, has kept his day job as owner of Wayne

Industries so he can Batmanize riddlers, jokers, and a few million people into paying six bucks for movie tickets.

Other multi-millionaires have chosen to put their money into more important things than destroying evil. A Saudi prince recently put up $500 million to bail out a near-bankrupt EuroDisney. In a year or two, Europeans will probably be going to Prince Al-Waleed bin Talal bin Abdulaziz al Saud Land.

Does anyone else think it's ironic that the Arabs, who have been taking over British businesses and New York taxis for over a decade now, have been invading French-owned businesses, long after France sent its Foreign Legion into Arab lands?

France has a minister of culture. (Possibly to monitor Arab takeovers of amusement parks?) Almost every country except the United States also has a minister of culture. This means either we don't have any culture or that politicians may think culture is what's grown in petri dishes. Either way, the lack of such a department says quite a lot about American values.

And speaking of American values, it seems that every time I turn on a TV set, some actor or actress is pumping the charity of the Moment. There's perennial favorite Jerry Lewis and Muscular Dystrophy. For the homeless, we have Billie Crystal, Whoopie Goldberg, and Robin Williams. There's outlaw country singer Willie Nelson for Farm Aid, and the squeaky clean Osmonds for the Miracle network. Even Sally Struthers gets into the picture by pumping for some starving foreign children who need our fifteen bucks a month so they can take correspondence courses and become brain surgeons. I wonder, however, with all the good charities taken up, if some aspiring actress was denied an audition because her agent couldn't find her a suitable charity.

Why does almost every starlet who's taken three acting lessons believe she has been given innate wisdom to "feel" her character and rewrite Arthur Miller's script?

Why do starlets say they'll do nude scenes "only if it's pertinent to the story," expose themselves in five centerfolds and 18 scenes in three movies, become mini-stars, then declare that because they believe in "family values," they won't do nude scenes no matter what the story line?

Why does the average person know who John Wayne Bobbitt, Tonya Harding, Joey Buttafuoco and almost every unclad starlet in Hollywood are, but have never heard of George Gershwin, Medgar Evars, or any of the members of the Supreme Court? And, can we somehow throw some of the blame to the media for pandering to our lowest basic instincts?

Why do TV interviews always have cutaway shots of the

reporter pretending to pay attention to what the news source is saying? Do we really need to see the reporter silently emoting three times in a ninety second interview?

Does anyone know why the TV camera focuses on the outfielder watching the homerun ball hit over his head, rather than on the batter who's running the bases? Couldn't those TV geniuses have figured out by now how to do a split screen?

I wonder if we'll ever see the time when late night talk shows book celebrity guests who aren't hustling a book, TV show, or new film.

And, while we're discussing TV, does anyone else get upset when newscasters pronounce nuclear as "Noo-kuhlar," February as "Feboo-ary," and library as "Lie-barry"?

Councils on literacy have done a wonderful job of creating flyers and bulletin boards to tell about their programs. But, if you're illiterate, how do you know what number to call?

Isn't anyone else offended that Pennsylvania's government and tourism industries have run a PR campaign to make the world believe, "America Starts Here"? The reality is that a better slogan might be "America Falls Into Potholes Here."

The U.S. Postal Service spends more than $70 million a year in advertising, including an ad in the upper right-hand corner of the Lifestyle and Sports sections of *USA Today*, for which a public affairs officer at the Postal Service claims he doesn't know how much it costs. Does anyone else wonder why the Post Office has to advertise? I mean, it's not like there's much competition for first class mail.

Why does it seem that small companies are trying to make us believe they're really large, with fancy titles for their staffs, while the megaconglomerates shower us with advertising they think will make us believe they're really our hometown neighbor?

Does anyone know why people pay $10-15 for T-shirts that advertise company names and slogans? Shouldn't the companies pay *us* to be their walking billboards?

The current fad in advertising is the semi-personalized telemarketers—"Hi, how you doing today?" Naturally, they're not selling anything, they just want to "take a poll" or do something nice that will improve my life. I wonder if we could take a lesson from corporate America and establish a system that tracks incoming calls and responds the way American business responds to us. Here's an idea. "You have reached the Schmidlapp Residence. If you are a bill collector, please press 1. If you wish to sell us a new and exciting product, please press 2. If you have a credit card you're hawking, please press 3. If you want to protect our credit cards from theft, press 4. If it's an extended warranty, just punch up number 5. For telephone companies that have 45 reasons why we should switch from our current phone service, your number is 6 . . ." We could set up an entire minute of menus, then when that poor fool falls for one, and punches the button, we'll just route him further. "You have pressed 2, telling us you're selling a new and improved product. If you have a product with wheels, press 1. . . ." If the telemarketer survives to the seventh sub-menu level and thinks he's found the right number, we just let him hear the final message, assuming he hasn't taken a pound of Prozac in the meantime—"You have punched the following numbers: 2, 3, 8, 5, 4, 3, and 9. Now, push No. 86." When he does, the phone disconnects.

Why is it that just about everyone believes he or she can become a writer part-time and earn goo-gobs of extra money, but when I say that I'm thinking of doing brain surgery part-time for spare cash, they all walk away from me?

By the way, has anyone else noticed that the best humor and social commentary columnists in the country all have last names beginning with "B"—Buchwald, Barry, Bombeck, Baker, Breslin and, of course, Brasch?

Finally, for the past five decades, millions of Americans have wondered how it was possible for the Holocaust to have occurred. We need look no further than America's reading habits for the answer. In 1993, the worst-selling single issues for *Time* and *Newsweek* were cover stories about the war in Bosnia. For *U.S. News & World Report*, the worst-selling cover story was about the famine and anarchy in Somalia. The best-selling issues? Cover stories about Rush Limbaugh and lesbianism.

America's Ding-a-Lings

America's telephone companies—they're the ones with the ding-a-lings—have been wallpapering the country with notices of how cost efficient they are. I have a friend who switches long distance phone companies every two or three months, staying just long enough to take advantage of the inducements to switch. She's long past the $25 free usage credit, and is now working on getting a two-week vacation in the Bahamas.

Switching long distance phone companies is relatively easy. Getting a phone in the first place isn't so simple, as I found out after moving. A few of the details are hazy, but I'm sure this is what happened.

"I'd like a phone," I said to the friendly smile.

"Ours or theirs."

"Ours or theirs what?"

"Ours or theirs telephone. For only $5.95 a month, you can rent a nice, beautiful, well cared for, always faithful telephone from us, or you can go to one of those cheap back alley stores that sell molded plastic they claim is a telephone, but is really nothing more than a fusion of leftover chemicals from Chernobyl." I chose theirs.

She asked me to pick a color. I asked for black. She told me candy apple red, tangerine, sunny yellow, lemonade, autumn gold, avocado, tropic green, white, cream, ivory, espresso brown, and 258 other colors no one but phone designers ever heard of, but no black.

I chose blue.

"Would that be electric blue, robins-egg blue, porcelain blue, or midnight blue?" I told her to surprise me. She did. "Would you now like to be able to call someone?"

"Doesn't my phone allow that to happen?" I naively asked.

"Oh, no, sir," she pleasantly informed me. "You need to call another phone company."

"But aren't you the same people?" I asked.

"WHAT!?" she screamed offended. "We're *not* the same people. We're different! We're our *own* monolithic monopoly!" So, I called another monopoly, a "Baby Bell" that gave me even more options and add-ons than come with a new car.

The first option I was given was to not have my name in the phone directory for which I had to pay the phone company a monthly charge not to do anything. During the next 11 hours, I also made decisions between rotary or pulse; chime, buzz, or gong; and call waiting, call forwarding, and call block, all of them leaving me call-pletely confused. But, at least I was assured of continuous uninterrupted service.

"If something goes wrong with my phone," I rhetorically stated, "I just call you."

"After figuring out where the problem is and how to fix it."

"If I knew what was wrong, why would I have to call you?"

"Because you have to know *who* to call. If it's in your telephone, you must call whoever sold you the phone. But, if it's in the inside wiring or any other problem, we'll come by and charge you forty or fifty bucks plus labor and maintenance."

"But, it'll be fixed, right?"

"Unless it's in your outside wiring. If that happens, you just call AT&T or some other long distance service provider, and they'll charge you another forty or fifty bucks plus labor and maintenance."

She took a quick breath, then continued. "Now, would you like our Reach Out Plan?" I wanted to reach out, but restrained myself, eventually learning that for a flat fee I could make an hour's worth of long distance calls anywhere in the country, except for long-distance calls in something I was told was a local long-distance area, which I doubt anyone understands.

It was finally time to add up the charges. She primed her computer, then began running the calculations—$4.75 for a dialtone, $3.50 for the federal line charge, $5.30 for local calling, $5.95 for the first phone rental, $11.90 for the other rentals, $3 for the colors, $1 for the first push button and $11 for the other buttons, a few bucks for product/service charges, a few more bucks for miscellaneous accessories, the state's relay surcharge, state and federal taxes, something for the 911 line, the $100 Deposit—"Did I also mention the installation charge? If we have to come out there, it'll cost you big-time, but if all we have to do is to throw a switch in our office, it's only $40. By the way, are you making a whole lot more since you moved?"

"What if I decide I don't want a phone?" I wearily asked.

"No problem," she said sweetly. "That'll only cost you ten bucks a month, and you'll still get an 18-page bill you won't understand at the end of the month."

JANUARY 1995

The Beeper Cacophony

It might have been an enjoyable party, but I didn't experience much of it since beepers were going off all evening, and all I heard were excuses of why used car salesmen, real estate agents, and grocery store clerks had to find a telephone.

"So, what's your sign?" asked a striking brunette who was beeped and never heard from again, at least by me. Apparently her sign was AT&T.

The knock-out redhead and I talked for five minutes before her beeper alerted her to call her service which relayed a call from her boss who wanted to know what color dress she was wearing to work the next day so he'd be able to color coordinate his staff. At least that's what I think she said, but I wasn't sure because she was beeped again.

The swaggering high school English teacher was beeped, but didn't have to run anywhere to find a telephone—he just unsheathed the one from his hip.

While waiting for a movie usher and waitress who simultaneously excused themselves to answer their pages, I overhead three people by the bar ask each other what our hostess must have been thinking to have actually invited someone so low on the prestige scale that he didn't have a beeper and car phone. "Could be a diversity thing," said one politically correct matron. "You know, we invite a Black and a beeperless columnist to our party."

Feeling alone and needing a drink, I asked the bartender for a virgin Piña Collada, but before he could crush the ice, he was beeped by Starbutt across the room who needed two whiskey sours with a twist of lemon.

After almost an hour of watching people flick off their pagers, and run to telephones, I noticed another soul all by himself.

"Party seems to be dragging," I said, opening the conversation.

"Yeah," he mumbled. "I just hope I get some action tonight."

"Since everyone's telephoning everyone else," I said, "I doubt there's much action anyway, especially when everyone seems to have blurred the lines between business and personal lives."

"It's now been 93 minutes, and no one has paged me," he said dejectedly. "It's so humiliating."

Not having done my good deed for the day, I sighed, and shuffled off to a telephone in the foyer. His pager beeped, and he rushed off to a phone about 15 feet away to chat with me about the price of kumquats. He was most thankful, especially when I didn't try to talk to him again so he could carry on simultaneous conversations with the striking brunette, the knock-out redhead, and some guy who was selling life insurance.

About the time I was ready to leave, the hostess told me I had a telephone call. It was Marshbaum wanting to know if I needed him to come in early the next day. "How'd you find me here?" I asked.

"I was driving along Route 11 finding dumb things for you to write about when I thought I should check in. So I called Horsehide who paged Littany who paged Bullnose who said you were at some muck-a-muck's party, so I called."

"*You* have a car phone?" I asked.

"Had to, Boss. Also a pager, CB unit, FAX, laptop computer, modem, and portable satellite dish. Gotta be on top of things in case you need a dumb idea at a moment's notice."

"When's the last time I needed you moments from deadline?" I asked.

"Makes no difference," he said. "*Sometime* you may, and you'll be happy you could get to me."

"That's all well and good, but *I* don't have any of those communications devices."

"Check your office in the morning, Boss. Got some nice units for you, too. It's only costing you a thousand or so a month to find me."

"Marshbaum!" I shouted, "I don't have an extra thousand a month to pay for cellular phones, paging equipment—"

"No problem, Boss. It's all tax deductible."

NOVEMBER 1993

The Impersonal Society

Some of my favorite people are the five ladies at the Bloomsburg, Pennsylvania, branch of the Philadelphia Federal Credit Union. Over the past decade or so, they have put up with a lot from me, with hardly an audible sigh, although I am sure there was a lot of cheering when my wife took over balancing the checkbook a few months ago.

The Credit Union ladies know my account numbers and status better than I do, have bailed me out of numerous problems, and have even gotten used to the reality that I know my accounts not as numbers but by colors. ("Could you please check my balance on Red savings, then transfer fifty bucks into Red checking, thirty into Blue checking, and a hundred to pay Blue loan?")

Even when they've had a long and tiring day, the ladies smile, joke, and ask questions about how I and my family are doing. The only thing they get from my "small potatoes" accounts is the occasional box of candy or a green plant and the satisfaction they're doing a good job, which doesn't even begin to add up to the personal attention they provide to keep my financial affairs in order.

But, during the past few years, because of a conspiracy by the grand poobahs of the urban corporate office who are into things like "time-management studies," the ladies have been slyly trying to convince me to use the push-button telephone to call an 800-number of a central computer where a digitized voice will tell me the status on my accounts, and allow me to electronically transfer funds from one account to another, and even pay bills. My human tellers even sweetly point out that it's easy to do telephone-to-computer transactions which are efficient and save me the $2 fee for almost every transaction a human teller has to do. I, of course, have just as sweetly explained that I have no ability to comprehend laborious written instructions that require me to push 132 different buttons just to hear a disembodied voice tell me my account is overdrawn.

With MAC drive-ups, direct deposit, and the phone, I don't ever need to talk to a human again. The reality is that I am willing to pay a penalty just so I *can* talk to a friendly voice in a rapidly increasing technologically imperfect impersonal society.

At one time, all telephone calls had to be made through a local operator who knew as much about you, your family, and the community as you did. Then, technology let us bypass a human, and do our own calling.

Call the average business and you now are greeted by a digitized voice giving you a menu. Listen to all the choices, push another button, and hear another menu. Some companies have four or five levels of menus, all so you can finally push a series of buttons and hear, "I'm sorry, I won't be in for the next six months. If you wish to leave a message, press 1; if you wish . . ."

We don't go to seamstresses anymore because we can now order by menu-driven telephone from the mail order department the same clothes everyone else is wearing. From vending machines, we can now buy not only candy bars and soft drinks, but insurance, VCR tapes, and even aspirin and condoms—and never have to talk to anyone. We speak into a clown squawkbox to order fast food which we eat in the car on the way to an aerobics class that treats us to a recorded cadence.

Although most clerks at supermarkets and department stores, who are usually paid minimum wage and receive no benefits, make at least an attempt to be friendly, an increasing number barely make eye contact while they languidly pass items past an electronic scanner.

With the computerization of America, you can now have your computer talk to other computers and make airline and hotel reservations, order furniture, get information from data bases instead of the library, and even hire a nanny, all without ever talking to a human.

On newspapers, we have replaced the wise older proofreaders and typesetters with dispassionate computers that have a passing knowledge of grammar and no knowledge of the community. Reporters are already researching and writing stories by calling up data bases, transmitting the finished product electronically to editors who send it electronically to the press—and no one has to talk with anyone else.

In California, neighbors stand next to each other before their fire-torn houses. For many, it's the first time they have talked to each other in weeks.

Even the lines that sound as if we care about each other—"I know where you're coming from," "I understand your hurt," and

"Thank you for sharing that," among dozens of others—are nothing but warm fuzzy codes so we can pretend we are communicating while we plan our next sentence.

It seems the only time we talk with each other is when we unite at sports events to shout "kill the umpire!" Most other communication seems to be flipping fingers and calling lawyers. Indeed, the Age of Communication has now become the Age of Uncommunication.

NOVEMBER 1993

Curbing the Paperless Explosion

It seemed like such a good idea at the time. A computer program that not only lets you fill in crossword puzzles, but also gives you hints, scores your results, and even lets you create your own puzzles. It was the perfect gift for my wife who lets a cup of coffee and the morning newspaper awaken her, then spends 10 or 15 minutes filling out the crossword as dessert.

But, a week after I gave her the super puzzles disk, she still hadn't installed it on the computer. Finally, she admitted that doing crosswords on a computer screen just wouldn't be the same as relaxing at the dining room table.

For most of us, newspapers, which have been around in one form or another for most of recorded history, just "feel" right. We can read them all at once or let them lie around, reading an article now and then wherever we have time.

We can cut out a picture, an article, or column, stick it on the Fridge or put a Post-it note on it and send to a friend. We can tear out an ad or coupon and use it later in the week. We can make printer's hats for our children, or take the day's news to the beach; when we're through reading it, we can put the newspaper over our faces to block out the sun. It's possible to take a portable computer to the beach, but when you put it over your face, you get real bad tan lines, and an even worse headache.

But, the Generation X-cess electronic whizzes and "mass communicologists"—who profoundly state that newspapers aren't even good enough to wrap fish or line bird cages anymore—claim the future is in the computer. They believe all knowledge can be encapsulated, then dished out at the user's preference. Interested only in the latest news about your favorite heavy metal band? Just type it in, then let the computer do the searching. Want only information about gardening? Bring up the right menu, spend another five minutes to select the appropriate submenu, then scroll through a list of titles until you find something interesting. Everything is available, and no self-serving gate-keeping editor is there to filter out 95 percent of the day's news. But editors help reduce the information overload that leaves most of us unable or unwilling to sort through piles of gigabytes to find what's important and what's fluff. Even newspapers and magazines have begun going "on-line" with stories and videos, figuring it'd be easier to make a buck off people who just want to read about the latest O.J. development than to understand all the news.

Actually, since large masses of the nonreading public seem to buy Sunday newspapers for the coupons anyway, it may be more logical to computerize everything. That way, readers can search for the right coupon—avoiding local, state, and international news—send an electronic signal to their grocery store, then pick up their discounted food whenever it is convenient.

With the local print newspaper, we see many things at once, subconsciously noting the size of headlines and the placement of stories; we get a "feel" for what is important, and how stories relate to each other. Even if we flip through each section to get to the horoscope and comics, we absorb some news along the way.

But, America's Chip-for-Brains have a response. With "pagination," they claim they can turn a screen into a miniature newspaper, complete with heads, pictures, and graphics. All we have to do is to try to read that full page on screen that's the same dimensions as a junk mail letter, move our mouses over what looks interesting, magnify the story, then scroll through it a few lines at a time until we need new glasses. Most of us can't even program a VCR, and these geniuses with their soft floppies really believe we'll become technologically sophisticated enough to nimbly wander through a maze of electronics and menus and actually "find" all the news that's fit to scroll?

Embedded within the computer age, the "futurologists" further argue that the death of print media is predetermined in an environmentally-friendly paperless world. At first, it seems as if the 25,000 60-foot trees we need to kill to produce the Sunday *New York Times* could be better used. But, almost all newsprint is now recyclable, and the Newspaper Association of America points out that forests are being grown specifically for newsprint production, significantly reducing the erosion of our existing forests and timberlands—and possibly stabilizing the rising cost of newsprint.

The reality is that newspapers will be around as long as advertisers determine it's profitable for them to continue to buy space on a piece of pulp paper 15-inches by 22-1/2 inches. When they get better results from handing out flyers in malls or running 30-second spots on "Gilligan's Island" reruns, then newspapers in print will no longer exist. Hopefully, advertisers will continue to recognize the reality of why print newspapers are important. We may complain about the local paper, but we still read it, each in our own way. So, whenever someone tells me that newspapers are dying, that in a decade or two they'll join the other dinosaurs of history, I tell them about my wife, her coffee, her voracious appetite to learn a little more about our world, her ad-clipping scissors—and her daily newspaper puzzles.

FEBRUARY 1995

Don't Despair, GM; Help Is on the Way!

It was probably the greatest outpouring of human warmth and compassion in the history of the American people. General Motors, the world's largest conglomerate, had just declared a $23.5 billion loss for last year, after suffering only a $4.5 billion loss the previous year. It was the largest loss in the history of any American corporation.

At corporate headquarters, thousands of concerned citizens came to help a fellow citizen who had fallen upon troubled times.

The owner of Bob's Service Station carried two bags of loose change for the ailing corporation. "All this time I was greedy," he sobbed, explaining how he kept charging full price to owners of GM cars that constantly needed work. "It was fun while it lasted," he said, "but my greed almost put GM out of business when all my customers started driving Toyotas and Volvos."

Charlton Pureheart, owner of Limited Used Cars, was also remorseful. "We probably cost GM a few sales because we were completely honest with our customers," he said. "We'll be more responsible this year and fall to the standards of honesty our profession is noted for."

"Although Americans have a competitive spirit," said Gretchen Schneider, a waitress from Fallen Arches, Illinois, "we were wrong to expect GM to have to compete against our allies, the Japanese and Germans."

Buckets Lowry, an unemployed steelworker from Johnstown, agreed. "Congress should have realized how hard it was for a company run by MBAs and lawyers to come up with original ideas, and should have forbidden all competition."

Among the crowd, vicious rumors flowed faster than gas through a four-barrel carburetor. GM executives were selling off their fleet of 747s for cheaper 727s; the vacation retreats would be closed part of the year; the executive dining rooms would now be opened to accommodate executives not earning at least a million a year.

Especially humbled were the newspaper reporters who now admitted their shame for having exposed GM's deficiencies in the marketplace. "We made fun of GM's stodginess and corporate stupidity," weeped the reporter for the *Daily Item*, "but we didn't know the people would actually stop buying cars that didn't work!" The reporter from the *News-Herald* remorsefully noted that he may have helped bring about GM's downfall when he wrote about cars that fall apart at 52,000 miles and trucks that explode upon impact, and that he probably drove the nail into GM's coffins-on-wheels when he reported that GM, sacrificing safety for appearances and cost-containment, recalled 1.5 million V-6 motors. Wiping a stream of tears, the automotive editor of the *Record-Standard* regretted his investigation that exposed GM as having fused its five car lines into one, changing only hood ornaments and decals to suit whatever preconceived prejudices the public had.

"We were too tough on them when we reported they didn't give any consideration to the people or the communities that cut them huge tax breaks," cried the reporter for the *Dispatch Chronicle*. "We should have applauded GM for closing 25 factories, laying off more than 100,000 workers, moving much of the production out of the country, and increasing executive salaries and stock options."

"Maybe we could find them jobs," one member of the crowd suggested.

"We tried that a few weeks ago," someone else said. "We took the executives to all the employment agencies, but found out they weren't qualified to be anything but GM executives or lawyers. It was very discouraging."

"Hey, Doc!" shouted a concerned citizen to a man just walking out of the headquarters, "how's it look?"

"Not good," said the physician, closing his wallet after making another visit to his ailing patient. "But, if we can keep the transfusion of money coming in, and don't demand any higher standards of production and community concern, then it's just possible we'll see these brave executives back in their corporate jets, and flying to vacation spots in the Caribbean."

Weeping in joy, the crowd rushed off to their banks to give even more to the company, while the reporters began churning out nothing but happy news. When last heard from, the patient was recovering, and looking forward to only a $20 billion loss this year.

FEBRUARY 1993

Mixing a Bitter Pill for America's Drug Companies

The advertising copy for the front of packages of Legatrin and Q-Vel are similar. The only major difference is that Ciba's Q-Vel uses capitals; and Columbia's Legatrin uses upper and lower letters. Beneath their logos, in small type, both companies proclaim the non-prescription drug is a "Muscle Relaxant/Pain Reliever." In bolder white letters, they also proclaim each "Prevents and Relieves Night Leg Cramps."

Not so, says the U.S. Food and Drug Administration. Not only don't the drugs prevent and relieve night leg cramps, but the quinine in the drugs can also cause severe medical problems, including visual, auditory, and gastrointestinal symptoms, as well as liver injuries, kidney failure, and severe dermatologic and blood damage. Hospitalization and deaths have occurred from reactions to the quinine, says the FDA which ordered the companies in 1995 to stop manufacturing and shipping OTC medicines containing quinine. Not affected is the use of quinine for treatment of malaria; because of almost insignificant doses, the FDA order also didn't apply to the use of quinine in beverages.

Columbia's Legatrin was the source of $4 million of its $9 million a year income; Ciba sold more than 1.3 billion cartons of Q-Vel between 1986 and its termination.

A century ago, Americans were unknowingly taking alcohol- and opium-laced drugs that were advertised as curing every medical problem from "indiscretions of youth" to syphilis and cancer. The patent medicine quacks were making $100 million a year from drugs that not only didn't cure anyone, but could lead to even more severe medical complications. Of that $100 million income, about $40 million went into bribing legislators to do nothing, and for newspapers and magazine editors to accept the ads and not investigate patent medicine claims.

In the U. S. Department of Agriculture, chemist Harvey Wiley spent 25 years trying to convince Congress that the patent "snake oil" medicines were harmful, only to be persecuted by his superiors. Finally, during the first five years of the twentieth century, several national magazines, risking their own financial health, refused to accept patent medicine advertising, and launched major investigations that laid out the facts behind all the "Wonder Drugs." But, the states' legislatures and the Congress refused to do anything, their allegiances paid for by business, pharmaceutical manufacturers, and by the medical establishment itself. Taking on the drug lobby, publisher William Randolph Hearst, then at the peak of his power, directed *Good Housekeeping* to establish a laboratory to investigate claims of all Hearst's advertisers, and to issue a "Seal of Approval" if the product did what it was meant to do without harming the public.

The Progressive administration of Teddy Roosevelt, spurred by the articles of several investigative journalists, had demanded passage of a pure foods and drug act. But, lobbyists had bottled it up in a Congressional committee. Then, in 1905, journalist Upton Sinclair went undercover in Chicago's meat packing houses to document corruption within the U.S. Department of Agriculture that allowed companies to package diseased meat. After most newspaper and magazine publishers, themselves subject to bribes and lobbying by the meat packing and drug industries, rejected Sinclair's investigation, a socialist company published the first of several articles that eventually led to the book-length publication of *The Jungle*, now regarded as one of the nation's most brutal novels. The novel so inflamed the country—causing Americans to boycott packed meat and foreign countries to refuse shipment of meat packed in America—that Congress had little choice except to pass the Pure Foods and Drug Act first proposed by Dr. Wiley.

Nine decades later, Americans may become upset with the high cost of drugs, but because of the FDA, we no longer worry about the safety of the medicines we buy from the neighborhood pharmacy; and, because of some courageous journalists, we aren't as suspicious of the advertising claims made by the pharmaceutical companies.

DECEMBER 1994

Selling Out America by the Yard

With the arrival of Spring, it's time once again to go through the house to find white elephants. Not the Caucasian males who are running Congress, but attic stuffers that can be cleaned up and sold at a yard sale. Here's a list of some of the better items, available at bargain basement prices.

• The first thing I'm getting rid of is my entire shelf of computer books. Maybe when the software companies figure out that clear, concise writing is the motherboard of effective communication, I may again buy some books. In the meantime, perhaps some needy family that can't afford dumbbells might stop by and purchase my collection.

• Also on the sale table is *How to be Your Own Lawyer*. After watching the Trial of the Millennium unfolding in an L.A. courtroom, I realize there's no need to have lawyers at all. Just plead guilty, serve the time, and you'll be out before the trial would have been over in the first place.

• One of my favorite books is *The Rights of the Worker*. But, in the present anti-labor mood of the country, this book is now categorized under fiction. It'll be available at a low take-it-away scab price.

Other items on the bargain table include:

• Two dozen diet books. The only thing I lost on this was a lot of time and money.

• Two dozen dog obedience books. Kashatten, one of the world's smartest German Shepherds, and I, whose intelligence remains in question, flunked out of not one but two dog obedience courses. But at least she's been able to get me to sit, heel, and roll over on her command.

• A sheet of apostrophes. This one should go quite fast since ads, flyers, and newspaper articles prove there's a desperate need for apostrophes.

• A case of White-Out. This also should go fast because ads, flyers, and newspaper articles also prove that when apostrophes are used, they're used incorrectly.

• Here's a 2-for-1 special. I once bought several dozen bumper stickers, "You've Got a Friend in Pennsylvania," hoping that by keeping the stickers off car bumpers, we might not give any neighboring states the impression that the Commonwealth doesn't know grammar. But then, Pennsylvania came out with an equally ridiculous slogan, "America Starts Here," and I just gave up.

• A 45 rpm record of "Happy Days Are Here Again," the Democrats' theme song, since there doesn't appear to be any Democrats left.

• A well-worn TV set, suitable for decoration or to hold plants. The set had survived 6,245 different talk-show hosts, the O.J. marathon, and "Thirtysomething." But, with deregulation by the Federal Communications Commission and the impending two-year take-over of the airwaves by the hordes of Republicans running for President, the set imploded.

• A 1,000 piece jigsaw puzzle of the Statue of Liberty. It once was assembled, but is now mostly in pieces. But, I'm sure someone out there, possibly a recent immigrant "yearning to breathe free," might wish to restore Lady Liberty before Congress deports him.

• A tattered copy of the Constitution, with part of the First Amendment missing. With the new Congress, I probably won't have any need for the rest of the Constitution anyhow.

The one thing I am saving is a copy of a piece of paper that says, "Governments are instituted among men, deriving their just powers from the consent of the governed. That whenever any form of government becomes destructive of these ends, it is the right of the people to alter or to abolish it, and to institute new government." The Communist Manifesto? No, the Declaration of Independence.

MAY 1995

The Day the Circus Came to Town

The circus came to town the other day, quietly and without fanfare. There was no mile-long parade with animals and bands, for the circus had snuck in on RVs and 18-wheel semis. No barefoot boys, their faces sunburned from the Summer, were idling by the gates hoping to be asked to water the elephants and camels in exchange for an admission. No one sold cotton candy or pink lemonade, for the menu was overpriced hotdogs, hamburgers, candy, heavily salted but unbuttered popcorn, and carbonated soft drinks, all of them huckstered throughout the show.

No one tried sneaking in under the main tent, for there wasn't one, just an impersonal green-tinged concrete-domed auditorium that also hosted farm shows and rock music acts. People no longer flocked to the sideshow because this time even the sideshow wasn't there—perhaps a tribute to mankind's sensitivity to human concerns, perhaps because the circus could still sell tickets without having to display society's "freaks."

The ringmaster, a hired actor wearing a tux—not even shiny black hip boots and a scarlet red coat—stood just outside the center ring, commanded the attention of his appreciative audience, and with a flourish blew his whistle and called out the acts.

First into the show was a trainer and some lions. The trainer snapped his whip, and fired a pistol loaded with blanks and special-effects smoke. The lions got up on stools, executed marching maneuvers—and yawned, undoubtedly wishing that the afternoon whistle would blow so they could quit work, go home, drink some beer and watch TV, or whatever it is that lions do when they're not working.

A couple of elephants and a half-dozen ponies ran around the center ring, stood up, bowed, and did tricks. All circus animals do tricks. Somehow, we believe that taking animals out of their natural environment, then forcing them to conform to what humans think is cute, is something to be applauded.

It costs a lot to feed and care for circus animals and the people who perform with them, rent auditoriums, buy newspaper ads, print tickets, hire a half-dozen local musicians, and pay the prodigious costs of liability insurance. In a wood-panelled board room in a steel-and-glass building far from the smell of sawdust, a group of directors, undoubtedly guided by an MBA CEO, advised by a gaggle of lawyers, none of whom would have the talent to do anything more than pick up after the elephants, may have determined that to maintain the price-to-earnings ratio, and to be able to continue to make quarterly dividend payments to its vaporous plethora of stockholders, it had to cut a few animals from the show, maybe also cut back on some acts and support staff.

Far above the grey concrete floor sprinkled with sawdust, above the safety net held by a dozen hefty roustabouts, swung the aerialists, their split-second movements drawing appreciative gasps. After them, an illusionist made doves appear and a tiger disappear.

Between acts came the clowns, a half-dozen of them prancing, chasing, falling to the syncopated rataplan of their own drummer, now throwing a bucket of water, now a bucket of confetti, blowing whistles and honking horns, bobbing and weaving and ducking. Once the highest form of comedy, clown comedy is dying. Instead of wearing fright wigs and red-bulb noses, baggy pants, floppy shoes, and polka-dotted stuffed shirts, today's upwardly mobile comedians wear designer jeans and theatrical makeup; instead of perfecting pratfalls, their managers help them perfect stock portfolios.

The show ended 90 minutes after it began. The actors returned for a mini-revue, then went to their RVs. The audience quietly walked out, past the entrance where they could still purchase circus T-shirts, programs, and recorded calliope music, assuming they didn't buy any during the dozen or so times vendors walked the stands just to make sure everyone who paid the eight buck admission didn't have anything left in their wallets or purses.

The acts that were seen only a couple of times in a lifetime now flood the television screens, and it's far easier at the end of a day just to stay home and push buttons on a remote control box. But the circus is our reminder of a childhood when we didn't worry about mortgages and the languid economy, when our lives weren't complicated by problems with the environment, street crime, or if we could scrape up enough money for an overpriced prescription. As with all things, we matured, and the circus matured, its freshness left for the young who will have their memories, and for the elderly who have everything to remember—and, maybe, for the dreamers who, like the clowns, will find their own drummers.

SEPTEMBER 1994

DATELINE: London Bridge, Arizona

In a way it's eerie seeing London Bridge standing in the middle of the channel dredged from the Colorado River, blending into its surroundings as if it had always belonged.

The stranger had come into town, and instead of becoming an outcast, took over, making things its way. It soon became so assimilated that almost no one now thinks it's odd that a 900-foot long granite bridge, completed in 1831 and which once spanned the Thames River in London, is out of place in the Arizona desert, in a place called Lake Havasu City.

By the bridge is a replica of an English Tudor village, the kind of village that kings and queens may once have visited. The stores, looking to the tourist dollar, sell numerous British-oriented items—desk flag sets with the flags of the United States and Great Britain, Shakespeare coloring books, British toys and toys that appear to be British, and all kinds of souvenirs from commemorative plates and mugs to British sailor caps and bumper stickers. In a remarkable bridge of friendship, a gesture of respect for the belief in the unity of all people, one shop sells not only British-oriented items, but American Indian wares as well—art prints and pottery, simulated leather purses, and even a rather cheap multi-feathered children's head-dress made in North Carolina.

There's a shop in which visitors may buy sculptured candles, each candle a work of art, the owner-artist having come from Massachusetts to the desert. Another shop features candy, not "ordinary" candy but rich chocolate imported from Italy. There's an English pub, still owned by the city of London, a geologist's shop where visitors may buy certified remnants of granite from the bridge, and a copper shop where tourists can choose among handmade plates, vases, and music boxes.

There's also the Drury Lane Theatre. Not the original, but a namesake where plays and movies from, and about, the British Isles entertain thousands each year. A red double-decker British bus, now embedded in cement, is where you can buy ice cream; nearby is a British cab, the background for thousands of amateur photos; a green freight handcart, also a picture backdrop, is in front of one of the stores.

In the park, with its thick carpet of deep green grass, in contrast to the rock gardens of most homes in the city, people can sit and relax and listen to the music of live bands at the base of the bridge. Beneath the first of the five arches is a concrete platform and stage where the city sponsors dances and various musical events, the shade of the arch protecting more than 500 people from the desert heat.

Under the bridge, beneath the other four arches, pass pleasure craft of almost every kind, from paddle boats to super-charged ski boats to houseboats, their passengers enjoying the cool blue water as it flows from the river, into a lake, then the channel, under the bridge, and past the English village.

Out front of the village, by the iron gates, by the metal lions on the concrete columns that guard the entrance, jesters give out announcements, their multi-colored costumes a stark contrast to the sunsuits and short-sleeve shirts of the tourists.

Further away from the park and village, from the bridge and its flagpoles which display American and British flags, is the downtown of Lake Havasu City with such appropriate businesses as the Shakespeare Motel, the Dorchester Inn and the Royal Inn. Even the Holiday Inn, separated from the lake by sand and a highway, appears to be a castle out of Camelot.

The newspaper that serves Lake Havasu City, as well as much of the Mojave-Colorado River area, is called *The Sun*. It's an appropriate name. The summer temperatures hover around 100 degrees, occasionally reaching 112, dropping into the 90s at night. Even in October, during the nine-day London Bridge Days celebration, the desert heat, still in the 90s, combines with a 25 percent humidity to sap the strength of even the hardiest visitor.

In the winter, as the temperature fluctuates between 40 and 80 degrees, the "snowbirds," people fleeing the snow and cold of northern Arizona and Colorado, descend upon the city. For six months each year, November through April, there are no apartments available to rent, and vacant hotel and motel rooms become scarce.

Although it's a concern to the year-round residents who fled urban sprawl, with urban problems, and who are now accustomed to the desert heat and a city that is isolated from any other city by at least 30 miles, the temporary growth each winter is accepted, just as the summer flood of boaters and campers and other visitors is accepted. The residents know there won't be many more permanent residents; few can tolerate the heat. The city that was designed for 80,000 residents has slightly more than 17,000.

Prior to 1964, there were no residents, no city and no bridge. Just a lot of sand and cacti and desert animals and, for a brief time during World War II, an Army Air Corps base. Then one day, Robert P. McCulloch of McCulloch Oil and McCulloch Chain Saw fame, flew over the site, looking for a testing spot. He liked the barren desert with the lake that had been formed when Parker Dam was created about 30 miles downstream in 1938.

McCulloch put up a hundred mobile homes for his workers, and the city was born. C.V. Wood Jr., president of McCulloch Oil, and a planner of Disneyland, became the planner for the city. Soon, McCulloch owned more than 16,000 acres of land and began to sell off the desert.

During the next few years, McCulloch flew in thousands of prospective investors and residents, giving them not only free transportation to and from Lake Havasu City, but a free meal and a sales pitch as well. Those who chose to invest in a concept of a planned city in the desert have prospered. But, there are those who called McCulloch a fraud—someone who bought land cheap and resorted to hucksterism to sell it off at a profit; they accused him of caring more for money than for people. There has never been a doubt that McCulloch, widely recognized as one of the nation's most creative businessmen-engineers, made money. But, in this age of cynicism and doubt, it's refreshing to know that everything Robert P. McCulloch promised was delivered. He told prospective residents that if they purchased a lot in the desert and built a house, not only would he pave a road to their door, he would also make sure that there were water and sewer connections, and electricity and telephone lines as well. The result has been a hodgepodge of streets, some forming half-circles and intersecting from other roads at two widely separated points, and with many streets with only one or two houses. But, McCulloch kept his promises. And when it appeared that the library would have to close, McCulloch put in enough money to make sure that it would remain a viable part of the community. And when people needed help, they didn't go to the Welfare Department; McCulloch was there with help and with jobs.

By 1968, there were 7,000 people living in the city; a homogenous population of residents fleeing urban sprawl. But these people were not yet the "desert rats" who grew up and would die in the desert. The city, and its people, were mocked, scorned, laughed at. It was difficult for a native Arizonan to believe that someone who grew up in the East's harsh winters, or the cultural climates of Los Angeles and New York, would adapt to the searing heat of Arizona, or the lack of what Arizonans thought that Easterners believed was "culture." McCulloch knew that something—he didn't know what, but *something*—was needed.

In London, the city government had determined that London Bridge, which had guarded the entrance to the city for 137 years—and whose predecessors had guarded the city for almost 1,800 years—was sinking. London Bridge was finally falling down, unable to sustain the weight of heavy traffic. McCulloch determined that the bridge—a bridge so tied into the culture of London—would be the focal point to bring respect to his city, and would also sustain it. And so the McCulloch Oil Corporation bid on the bridge—in typical McCulloch fashion. First, the company reasoned that if the city sold the granite from the bridge, it would earn $1.2 million. The corporation doubled that amount, then added $60,000, $1,000 for each year that McCulloch was alive. For $2,460,000, Lake Havasu City had a bridge.

Each granite block was carefully removed from the bridge, numbered, and shipped to Long Beach, California, then trucked to the desert. In one of the great feats of engineering and stone masonry, and under the direction of British engineer Robert Beresford, London Bridge, with its five arches and British lamp posts, was reconstructed on sand mounds in the middle of the desert. Steel and concrete provided a stronger inner foundation—and a lighter bridge since engineers determined that a bridge entirely of granite would sink into Arizona's sand, much like it would sink into the London's Thames had it remained in London.

On October 10, 1971, three years after the first stone was

removed in London, and a one-mile channel having been dredged so that water could be diverted from Lake Havasu and the Colorado River, London Bridge was formally dedicated. Robert P. McCulloch, wearing light-colored slacks, a dark blazer and a striped tie, stood next to Sir Peter Studd, the Lord Mayor of London, wearing a ceremonial robe with medieval ornateness, and smiled broadly as thousands of balloons were set free. A dream had become a reality.

MAY 1983

Disneyland: Where Fantasies Are Real

A little girl—perhaps three-years old, no more than four—stood at Carefree Corner and watched the parade go by. The pupils of her eyes were wide open, for she wasn't about to miss anything. Not the pretty floats that showed the nation's history. And not the eight-foot tall characters with their colorful costumes and oversized doll-like papermachè heads, imitations of the people who shaped America's history.

Beside her, tightly clutching her left hand, was her grandfather, his hair now gray, his face covered by the years of history, a gentle man remembering parades of long ago. Together, they were at Disneyland.

Disneyland—Walt Disney's magical kingdom, the kingdom where fantasies become real, if only for a few hours. Disneyland—with Peter Pan, Snow White, and Alice in Wonderland rides to be enjoyed by the little girl and her grandfather. Disneyland—with the Mad Tea Party, King Arthur Carousel, Flying Dumbo, and a 146-foot replica of the Matterhorn, complete with mountain climbers and bobsleds. Disneyland—with the overpriced concessions, and myriad guides and guards, all seemingly with the same slim bodies, interchangeable faces and short hair styles.

And, now, almost in front of the little girl—so close that she could almost reach out and touch them if there were not cords of rope that kept her back—were Mickey Mouse, Goofy, and Donald Duck. The three of them in patriotic costumes and resembling the flutist, the drummer, and the wounded soldier in the painting, "The Spirit of '76." This was 1976, the year of the bicentennial.

Walt Disney was an innovator; a man whose imagination made things happen. It was his imagination that led to "Snow White," the first full-length animated feature. It was his guidance, backed by his brother Roy's financial and administrative wizardry that had led to many of the standard innovations in animation.

And, if at times it seemed that Disney had formed his own ideas, "blocked in" his thinking and forced his staff, perhaps the most creative staff in the industry, to accept his, and only his ideas, few complained.

Those who worked for him—those men of genius with the names Dick Huemer, Wilfred Jackson, Ward Kimball, Ub Iwerks, Cal Howard, Jack Hannah, Ben Sharpsteen, John Hench, and others whose names have little meaning outside the industry—willingly accepted a few restrictions and a place away from the footlights in order to be with Walt Disney and his operation.

For many years, Walt Disney had a vision. A vision of a land where the problems of today could be forgotten, a place where people could be happy and enjoy life. On almost 200 acres of land near Los Angeles, land once covered with orange groves, Walt Disney decided to build his dream.

The orange groves came down, but almost a million plants, shrubs, and trees would replace them. On July 17, 1955—a year and a day after ground breaking—Disneyland, the world's largest amusement park, was dedicated. It cost $17 million, and had 22 major attractions in five theme areas. There was Main Street, a reflection of an exciting town at the turn of the century; perhaps, the Kansas City area where Walt Disney grew up. And, there were Fantasyland, Tomorrowland, Adventureland and Frontierland, all leading into the Storybook Castle in the center.

In the year of the bicentennial, 21 years after the dedication, the value of Disneyland is $150-million. There are now 50 major attractions. And, there are two more lands—New Orleans Square and Bear Country.

In the coming years, there will be even more attractions, as Walt Disney once said that Disneyland would never be completed. During its 21 years, more than 140 million admissions tickets have been collected. With the millions of dollars in taxes Disneyland has paid to Anaheim, the city grew and expanded. In 1955, there were only about 23,000 people living in the city; there were only 34 restaurants and motels.

Two decades later, there are more than 180,000 residents; 275 restaurants, and 125 hotels and motels with more than 10,000 rooms, including the 929 rooms of the Disneyland Hotel.

With the taxes, the city of Anaheim built a new library, new governmental offices, and larger recreational facilities for its residents. It even attracted a major league baseball team, the Anaheim Angels who later changed their name, but not their city, to the California Angels.

The little girl didn't know any of this. Perhaps she never will. For her, history is what she was seeing now—in front of her.

Another float. Then another. Each float bearing happy characters from a part of America's history—*her* history.

And then a band. Soldiers of the Revolutionary War, their uniforms red and white, playing fifes and drums, marched before her. The little girl saw the soldier-musicians—White soldiers marching with Black soldiers, conspicuous by their presence.

Throughout the parade, Black and White characters marched together, integrated within the nation's history. The violence and mental torment that the nation had directed against the Jews and the Blacks, against the Irish, Chinese, and all the other minorities, is not a part of this parade. This parade is a happy parade; this parade does not reflect history the way it really was, but the way it should have been.

The little girl was too young to know. She was too young to understand why some of the lyrics in one of the songs in "America Sings," one of the major attractions at Disneyland were changed because some people thought the lyrics were racist.

And the little girl was too young to know that there is still much prejudice in the world against the Jews . . . and the Blacks . . . and the Chicanos . . . and against any minority group.

Now, in front of her was a keel boat, its yellow trim on brown wood brightly reflecting the sun. The keel boat in the parade moved on wheels; its passengers, Black and White, in brightly-colored costumes of the early decades of America's history, were Disneyland employees.

Over in Frontierland, where the nineteenth century is "today," two keel boats—"Mississippi mudwumpers," according to the Disney people—take paying passengers on a mile-long river tour around Tom Sawyer Island.

Also in the river, and also carrying paying passengers on the 8 to 13-minute journeys, are island rafts, canoes, a full-size reproduction of the "Columbia," which in 1790 became the first American ship to travel around the world; and the "Mark Twain," a three-deck paddlewheel steamboat patterned after the great showboats that once traveled the Mississippi.

On the island itself is a child's wilderness fantasy, complete with a forest that leads to a fort, a treehouse, an old mill, a cave, and suspension and barrel bridges—bridges to be climbed on, not looked at.

In Adventureland, it's possible to take a jungle cruise aboard a colorful launch. The launch, moving on hidden tracks, passes tribal villages and ferocious animals, all of them appearing to be real, but all just part of the magic of electronics and carpentry and inaudible impulse signals of machines that program their every action.

At the same spot, on every cruise, a hippopotamus will rise out of the water and threaten to overturn the launch, only to be shot back into the water by a courageous boat skipper, his nine-minute narration—"Watch out! To your left we have a charger—he's coming at us!"—carefully rehearsed to appear natural and spontaneous.

In Fantasyland, you can pilot your own motorboat or take a Storybook Land cruise on Dutch canal boats that will pass through the mouth of a whale and past green hillsides that cradle miniaturized sets from many of Disney's animated motion pictures. And, in Tomorrowland, the eight "nuclear" submarines, powered by electro-diesel engines, will show you an undersea world of "liquid space."

In Disneyland's transportation corps are turn-of-the-century fire engines, horseless carriages, double-decker buses, and street cars.

And there are trains, something Walt Disney himself had a special love for. Here, in a world of fantasy, trains take passengers through three-dimensional dioramas of the Grand Canyon and a Primeval Forest on their one and a half-mile journey around the perimeter of the park. They also take passengers through a mine and into a wonderland of the Southwest or on a "circus trip." Overhead, the world's first functioning monorail, takes passengers from Tomorrowland to the Disneyland Hotel and back, a 2-1/2 mile lap.

And now in front of the little girl are two more trains. But, these trains were made not from steel, but from wood. The two trains, their cowcatchers facing each other as they moved past the Castle and near Carefree Corner, showed a united nation, a nation once torn apart by the Civil War but now brought together by the first transcontinental railroad.

For a contemporary world torn apart, the people of Walt Disney had produced "It's a Small World," a subtle statement that among the children of the world there are common bonds that should unite, not destroy people.

At the New York World's Fair in 1964, where the attraction premiered before being moved to Disneyland in 1966, people waited as much as two hours for a few minutes' journey by boat through a maze of channels in a darkened auditorium in which children-like dolls, each moving by predetermined electronic direction, sang in harmony that the world wasn't all that big; that it was really "a small world after all."

For the little girl, her world—the past and the future was here—and now—in an enchanted kingdom larger than she had ever known. By the end of the day, as the lights of main Street showed their fluorescence, the little girl would become tired, perhaps a bit cranky.

She might have seen, and perhaps understood, the meaning of the Small World attraction. She might have seen ghosts in a haunted mansion, pirates on a burning ship, and Indians in the forest. She might have seen an owl and an eagle host a musical tribute to America or 255 electronic birds, perform an international stage show, or even meet a giant squid beneath the ocean.

She might have gone on a mission to Mars, traveled through the nucleus of an atom, and flown through Peter Pan's Neverland. She might have ridden an antique fire engine, a wooden horse on a carousel, and one of the island rafts. She might have eaten more than she was supposed to, walked longer than she ever did, and laughed, cried, screamed, smiled, and enjoyed life perhaps more than she ever had before.

She would go home that evening where another real world exists, a world that may not be so nice.

Since 1955, when Disneyland was dedicated, there have been thousands of stories about this enchanted magical kingdom with life-like animals, unusual rides, and happy people. A few cynical journalists went "behind the scenes" to "investigate" Disneyland—much in the same way they would "investigate" any government. And, as with any government, they found flaws.

The war in Vietnam had shocked America out of an age of complacency and into an age of cynicism and "relevance." For many, Disneyland wasn't "relevant" to the world "out there." Some have called it a "rip-off," with its overpriced hamburgers and fries that wouldn't rival even the fastest of fast food restaurants.

At the entrance to Disneyland is a sign, perhaps missed by many as they enter Main Street and a world of 1900. That sign, a statement of purpose, simply says, "Here you leave today and enter the World of yesterday, tomorrow and fantasy."

If Walt Disney had never lived, and if there was no Mickey Mouse, there will still be a Disneyland. It would have been called something else; it would have been different.

But it—something—would be there, for people need to escape, to enter another world—a "different" world, even if for only a short time.

JULY 1976

For Casinos, 'Six-Hour Visitors' Make a Full House

She's a 77-year-old slightly overweight white-haired widowed grandmother from northeastern Pennsylvania who has worked the past 15 years for the state at a minimum wage "Green Thumb" job. But five or six times a year, she and two of her closest friends pay $22 to a bus company that hauls passengers the four hours to Atlantic City.

"I can't afford *not* to go," Gloria Adams says with a twinkle. Depending upon which casino the bus goes to, and which day of the week it is, each bus rider receives $10-$15 in tokens from the casino, usually another $3-5 good for the next trip, and a $3-5 meal credit. She can choose from any of 92 restaurants or 30 cocktail lounges; if she eats at one of the lower-priced buffets, her meal is almost free. To lure 31 million "visitor trips" a year, one-third of whom come by bus, casinos spend about $800 million a year, one-fourth of their revenue, on promotion.

For most senior citizens, the $15-25 worth of "comps" is sufficient. But, for the late evening "high-rollers," the ones who think nothing of dropping a few hundred every Black Jack hand or roll of the craps dice, the casinos will provide free limousine or helicopter service, luxury suites, food, beverages, show tickets, myriad trinkets from keychains to suits, and just about anything a loser could want.

Every day, about 800 buses, each carrying 40-50 passengers, most of them women over 55 years old, pull into the casinos, stay six hours, then leave. With travel time, combined with limited bus parking, the casinos have figured that six hours is the "right" time for the "low-rollers." On the casino floors the senior citizens scramble to their favorite slots—they have 23,000 to choose from—and most won't leave for three or four hours, just shoving coin after coin into the machines, watching the dials spin, and hoping for a jackpot. They'll crowd the nickel slots first; few ever try one of the 46 $100 slots. They'll put in five coins, get a cherry, and win two coins which hit the metal coin tray and boldly announce yet another "winner." Sometimes, they'll get mini-jackpots of five, ten, even twenty times what they shoved into the slot. But, they'll recycle the change, hoping for the Slotbusters payoff that'll give them and their grandchildren an income for life. Sometimes, even if there are people waiting, they'll play two slots at once; if one doesn't pay, its neighbor will—at least that's what many figure. And most go home with less money than they came with. It's not unusual for bus travelers to drop a hundred, two hundred dollars in their allotted six hours at the casino. A few have dropped most of their entire month's social security checks.

For their part, the casinos tell their players—in TV commercials, on all the printed literature, even in random announcements beamed from concealed loudspeakers—to "bet with your head, not over it." But then they put an acre of machines, tables, dealers, and provocatively-dressed hostesses into an air-conditioned room that has no windows or clocks, and entice their victims to sign up for plastic card memberships that, when inserted into slots, record the number of times a coin was dropped, and lead to even more trinkets to lure them back into the casino. The casino also sends employees around to convert currency into coins, and provides free drinks for parched throats that yell encouragement to the machines, and "wipes" for fingers that become dirty handling all those winning coins. For those who have arthritis or don't have the energy to pull a lever, the slots even have push buttons. The casinos want their victims to sit there, in one spot, and lose. If they could figure out a way to catheterize the players so they don't even have to go to the bathroom, they would.

Gloria Adams often comes back with more than she began with. "I know my limits, and I don't go over them," she says, emphasizing that unlike some players, she never resorts to using a MAC card to get "just a little more" playing cash, just in case the "big one" is on the next slot pull. It doesn't upset the casinos. They like it when people win. Nothing entices other people to lose money like the ching-chang of coins dropping into slot trays, and the excited boasts of winners back home. But, most of the gamblers are supposed to spend the freebies, then gamble some more. It has worked very well for more than fifteen years.

According to the Casino Control Commission, the "handle," the amount of money bet, since the first casino opened is about $30 billion, about 55 percent of it from slots. Last year alone, the revenue, the amount of profit prior to expenses, was about $3.2 billion. By law, each slot machine must pay back at least 83 percent of its take; most now pay about 90 percent. The Casino Control Commission even makes public which casinos have better payoffs and which groups of machines pay better than the others. But, few study the numbers, and none of the casinos see any reason to post them. With accountants taking advantage of every tax law and loophole, and with the casinos themselves heavily leveraged, the after-expenses "net" last year was a mere $165 million.

The first of the twelve casinos went into business in 1978 following a statewide referendum to allow gambling in a city that had combined Miss America, salt water taffy, rolling wicker chairs, a four-mile boardwalk—and a deteriorating downtown, tenement housing, and the state's highest violent crime and poverty rates. Gambling will save the city, proclaimed the voters, the legislators, and even businessmen who planned to mine the east coast, smugly knowing that a third of the nation's population live within six hours driving time of Atlantic City and could easily be lured to a gambling empire that promised instant wealth. For its part, the casinos since 1975 have paid over $2.5 billion in revenue taxes, $850 million in property taxes, $575 million in regulatory fees and licenses, $760 in federal corporate taxes and $250 in state corporate taxes. Another $375 million, $40 million last year alone, from a 1.25 percent tax, went to Atlantic City's Casino Reinvestment Development Authority to be used for civic improvement.

For its part, the state of New Jersey has been so appreciative of receiving money, it built a 43-mile long expressway from Philadelphia directly into the Boardwalk. To lure even more middle- and upper-class marks to the casinos, the city and state are building a $254 million 500,000 square floor convention center and a $70 million three-block "corridor" to link the center with the Boardwalk.

By the mid-1980s, speculators and the casinos had gobbled up all available land in Atlantic City, expecting as many as thirty-five casinos to rescue the city. Only when it appeared there would be a saturation of twelve casinos could the school district (which last year received 69 percent of its $33.5 million property tax income from the casinos) finally buy enough land to relocate its high school. But, the taxpayers of the proposed $80 million high school still had to pay $8.7 million for 48.8 acres, a lot for a school district but less than the $4.6 million the Showboat recently spent for three acres so it could add 200 more hotel rooms.

Although city officials are now proudly proclaiming a renaissance for the ocean front resort, which has received about $500 million in property taxes from the casinos since 1978, about 69 percent of its property tax base, they are also faced by the blemish that four of the past six mayors were indicted on a variety of malfeasance and criminal charges, the downtown has deteriorated even further, population has declined 14 percent since the first casino opened, there has been an increase in violent crimes and prostitution—and poverty is still the way of life for a large chunk of the 38,000 residents who live in a city with no theaters and only one supermarket. So deteriorated has Atlantic City become, even with its proposed massive revitalization, that most of the 41,800 casino employees won't even live in the city where they work.

Nevertheless, no matter what the city's problems are, the buses still come to Atlantic City, and senior citizens flood the casino floors by day, while the high-rollers take over by night. And the ching-chang of coins continues to lull Americans into believing that the good dream belongs to them.

"Gloria Adams" is a fictional name for the lady who asked that she not be identified.
AUGUST 1993

Conventional Fund Spending

In the same month Philadelphia opened its $522 million 1.3 million square foot convention center, the most expensive public works project in the city's history, the Philadelphia school district ended its year having had to make $60 million in personnel and program cuts.

The civic center cheerleaders cite numbers to show how beneficial the center is to the people. The average attendee at one of the already-booked 140 trade shows will stay three nights in Philadelphia and leave $1,040, the average attendee for one of the already-booked 120 conventions will leave behind $688, and even the locals, defined by the center as "leisure travelers," will each leave behind $75 when they attend a trade show, convention, or meeting, according to the International Association of Convention and Visitor Bureaus. The economic impact by the groups already booked is more than $350 million, and if the tentative bookings turn permanent, the impact could be as much as $616 million, we are told. Even the 1,200-room Marriott Hotel already has 80,000 reservations for the next decade.

The Center says 6,000 new permanent jobs, about 80 percent of them in the lower-income service industry, will be created by the end of the century. And, they proudly point to a $10 million decade-long education program that will train persons in four to eight week programs to be entry-level service industry employees who will become the chambermaids, waitresses, cooks, and desk clerks at the center and hotel to assure a permanent low-income staff.

And, say the cheerleaders, most of the $522 million cost of the center will hardly cost anyone anything since the state provided $185 million of the cost, the city floated bonds for another $277 million, and the rest is going to be repaid from hotel occupancy taxes.

Here's a few more numbers that the cheerleaders don't mention. Had the 3,000 persons who attended a $250 per person black-tie opening night dinner, and the 10,000 who attended a $25 "FunConventional Fling" decided to stay home and donate the $1 million cost of their tickets to the school district, quite a few teacher jobs could have been saved. Had just half the $2 million spent for artwork in the new center been donated to the school district, creative arts programs could have been significantly expanded. Had just half the $3.6 million 11-day opening promotion budget ($1.1 million provided by the state) been given to fund cancer research at the Temple University medical school, perhaps we would have a major advance by this time a year from now.

Had the state's $185 million construction share been used for existing state programs, it would have funded about half the annual budget of the State System of Higher Education or two-thirds of the PennDOT budget. It could have funded almost the entire health department or Department of Environmental Resources budgets. At today's costs, it might have funded any of the following—the state police for almost two years, the labor and industry department for the next four years, the state's share of helping the homeless for almost 12 years, the literacy program for 26 years, or either the emergency management program or the Scranton School for the Deaf for the next 37 years. Or, it could have been used to help improve the lives and programs for the elderly, disabled, and homeless.

Just two years ago, Philadelphia was on the verge of bankruptcy. It responded by slashing budgets from all departments, and severely cutting back on libraries, fire stations, and homeless shelters. Those cuts have not been restored. Had Philadelphia's share been spent on the 5,000 homeless, half of them in the city's excellent shelter program that sustained heavy financial cuts, perhaps there wouldn't even be any homeless today.

In Atlantic City, an hour by expressway from Philadelphia, the state of New Jersey is building a $254 million 500,000 square-foot convention center that will sit at the base of a city-financed $70 million three-block corridor to the Boardwalk that the city hopes will revitalize a decaying downtown. But, the Center and corridor will primarily help corporations, conventions, and maybe the 31 million "visitor trips" the casinos will record this year. However, even with the city receiving an annual jackpot, poverty and urban blight, marked by gutted stores and abandoned houses, stretch from Route 87 to the Boardwalk.

Seattle completed a $158 convention center in 1988. Five years later, Boeing, the largest employer in the city, is in the process of laying off 38,000 workers, 25,000 of them in Seattle. In Manhattan, which has 49.4 million square feet of vacant office space, the $486 million Javits Center, with 720,000 square feet of exhibition space, is only seven years old.

During the past two years, convention centers were built on South Padre Island, Texas ($10.5 million); Moline, Illinois ($33.4 million); Austin, Texas ($50.4 million); and Columbus, Ohio ($94 million). In planning or construction are a $52 million river-front convention center for Mobile, Alabama; a $141 million convention center for Charlotte, N. C.; and a $290 million hotel-parking garage-convention center in Providence, Rhode Island. Palm Beach, Florida, is thinking about a center. We are told these centers benefit all the people. But, the homeless, unemployed, and low-income Americans usually don't attend conventions.

Our environment reeks of decades of abuse, we still haven't found cures to the health care crisis, more than 8.8 million Americans are unemployed, another 250,000 are homeless, and almost 25 million low-income Americans, about one-tenth of our population, are receiving food stamps. The money spent for convention centers could be spent on a massive jobs program that trains people not for "service" jobs that benefit the few, but for entry-level training in the health professions and environmental planning, social work, transportation, and labor education. But we're building convention centers. Something just doesn't seem right.

JULY 1993

The Monday Night Football Game

The minuscule cocktail tables, red cloths draping their round tops, threaten to tip when pushed; nevertheless, they serve their dual functions as repositories for drinks and cigarettes, and as excuses to cluster chairs.

The semi-stuffed chairs that surround the tables, however, are sturdy, made with a wood-like veneer for the back and base, and covered with a synthetic something-or-the other that is supposed to look like leather. At the beginning of every evening, four chairs surround each table, making for intimate experiences, but the chairs roll, and during the evening will roll back and forth from table to table.

Along one wall, with its fake fireplace motif, are recently-manufactured Victorian-style couches and high-backed chairs. On the walls are photomurals—one of deer grazing in the field, the other of geese in flight. Near the geese are fake-wood bookshelves supporting *Reader's Digest* condensed books and plants that will never die.

A sixty-foot U-shaped bar edged in naugahyde softness, guarded by 25 or 30 fake-oak $80 bar stools, is on the opposite side. Against another wall, not far from wood-panelled restrooms, six mikes and four speakers quietly frame a Lilliputian bandstand that supports a set of drums and an electronic organ programmed so that anyone who can count to 4 can play it. On Thursdays, Fridays, and Saturdays, three singer-musicians will have their sounds amplified to the threshold of maybe wilting the plastic flowers.

Wood beams and amber chandeliers above, red and black patterned carpeting below. The lights are forever dim. Dim for effect; dim for illusion.

It's now 6 p.m., Monday. There are a few people here, most in their 20s and 30s. At the bar sits a lady, perhaps 25 or 30, maybe more, flanked by two men. The lady could be a secretary or junior executive. Her golden highlighted hair is jelled, curling-ironed, permed, and blow-dried to give her a perfectly natural look. She wears makeup and lipstick; Max Factor on her face, Revlon on her lips. Her fashionably low-heeled pumps add an inch to her height. Her pink silk blouse and tan wool skirt could not have been bought more than a month ago. The two men, also in their late 20s or early 30s, could be lawyers or junior executives or ad salesmen. Suits of blues, ties of red. Modishly-long razor-cut blown-dry hair, as painfully attained as the lady's, covers half their ears. A slight fashion statement, but nothing too extreme. Clean-shaven with a gentle splash of Bijam. Expensive-looking watches and 10-karat gold college rings.

First one, then another, talks to the lady. Then both talk to her. For part of the evening, she'll delight in being in the middle. They'll talk to her and with her; maybe later, about her. But right now, they're both talking with her. About nothing really. They had earlier identified themselves by their occupations; she had identified herself by her astrological sign. Of course, this had meant that the two men had to identify their own signs. Compatibility is important in a place like this.

A man sits nearby, a touch of talcum and after-shave hinting the air, a casual synthetic drip-dry V-neck hugging his rock hard chest and abdomen, drip-dry 501s hugging his waist and thighs, imitation leather shoes caressing feet that have been adequately powdered. A few moments later, another man, almost a copy, but with a golden chain dangling in his "V"—few men will be bare-necked—sits down. There's now one immaculate lady surrounded by four *GQ*-approved men. To others, it appears that they're close friends—buddies—pals. Singularly, they talk with her. About cars. About careers. They talk about this. And they talk about that. About promotions and job security, racquetball, jogging, or time-sharing. About something each of them read in the *National Enquirer* or *Penthouse*, but which they remember as having read in *The Wall Street Journal*, *Esquire*, or *Business Week*. The high cost of apartments and condos is a major problem. Finding a good stereo system or cellular phone for the BMW or RX-7 is a major problem. Every now and then, there might be a discussion, but never controversy, about the something in the news, their mouths parroting *Time* or *Newsweek*, their concern as deep as cocktail glasses, as lasting as cocktail napkins.

Cocktail waitresses—costumed in black high heels, black-net hose, micro-mini fluffy skirts, push-me-up bras disguised beneath titillating-cut blouses—hustle the drinks while the customers hustle each other. No glass will be allowed to be empty for more than a moment. There's always another drink to be served, another tip to be earned.

At a table, a young man orders a Vodka-7; at another table, a young lady orders a Vodka-7. They have something in common. They glance at each other, and away from each other. She gets up, goes to the *hors d'oeuvre* table, and delicately places small carrot sticks and shrimp puffs onto a small paper plate. It's just enough, but not too much. Must watch that waist. She turns, and almost falls over the gentleman who exchanged glances with her. She apologizes. He apologizes. They apologize. They giggle over their embarrassment of almost falling into each other, begin some small talk, then wander to a table at a neutral site. It's all so superficial, but no one cares. They need to hustle so they can be with someone. So, they'll keep searching . . . reflecting . . . being conversationally attractive.

A pair of "barely-21s," a few months into the work force, enter and sit at a table. She's carded, her ID discretely checked by a waitress. This is too lucrative a business to be jeopardized by a Liquor Board violation. Her companion, blowing smoke rings from a Marlboro positioned beneath an anemic brown-blonde mustache, is accepted without ID. They order piña coladas and think they're in love.

Two 30ish ladies sit near one of the bookshelves. They talk what must seem to be "girl talk." Two 30ish guys sit near another bookshelf, talking "man talk." It isn't long before there are four people at one of the tables and no one at the other, and God-only-knows what they have to talk about.

The regulars greet each other; the infrequents settle in—watching—waiting. Four more razor-cut, blown-dry junior somethings enter, survey the swingles scene, find everyone attached to someone else, order Heinekins and Bud, talk, joke, make polite noises, and imagine they're successfully cool.

Near one of the photomurals sits a lady with a vacant sadness in her eyes as she sits alone, reflecting—thinking—maybe waiting. For a friend? Lover? Nevertheless, she's attractive, and probably won't be alone long. Someone will claim to have just the same values she has—and wouldn't she like to go somewhere else?

It's now 7:30. More than a hundred people are here. Talking. Chatting. Discussing. Talking about jobs and vacations, cars and condos. Every day it's the same. Every day the people are the same. And they all talk about never—*ever*—planning to get into a rut like Jim or Karen or Roger or Maureen. They all reaffirm that although they love their jobs and are more than amply rewarded, soon they'll be promoted or move to something better. There's always something better, whether people or jobs. And they're so busy hearing themselves that they don't know that no one's listening, but they'll be sure to "thank you for sharing that," or letting us know that they "can relate to that."

For the last time this evening, the *hors d'oeuvre* table is replenished, and the price of drinks goes up another fifty cents; Cokes and beer are now a buck-fifty, mixed drinks as much as four bucks. The sandwich-and-salad special is $7.95—chips and pickle slices included.

Nine O'Clock. The bartender pushes buttons on a remote control, and from a large-screen television, Eagles and Giants appear. There's noise from the television and noise from the lounge. People are chatting and watching, hustling, munching, and drinking; in their confusion of who they are, and what they want, they'll settle for instant friendships made in front of a cathode ray tube that changes pictures 30 times a second. The Monday Night Football Game is about to begin.

SEPTEMBER 1995

Sammy the Hustler

He walks into a bar, sits down and orders a beer, not because he likes beer but because he wants to blend in, to be "just one of the guys." He's wearing jeans, boots, and a red flannel shirt tonight; sometimes he wears a suit; sometimes jogging sweats. It all depends upon the bar, although he prefers places where the young professionals flock. Once, he earned $3,860 when he accidentally stumbled into a bar full of half-drunk conventioneering lawyers who were just a little too stupid, and a whole lot greedy. In mankind's determination to prove human superiority over the machine, Sammy has found his career; the pinball is his weapon, greed his accomplice.

Sammy is a hustler.

It makes no difference what bar he's in, just as long as it has at least a couple of pinball machines and no video games. Sammy hates video games. Functionally illiterate fifth graders can zap alien invaders from TV screens all afternoon for just a quarter; it's not real, not like the three-dimensional world of a pinball machine.

He walks over to the machines, and watches the players who are determined that steel balls shall hit the bumpers and eternally be flicked back into play by the flippers. No matter how many points the player racks up, it's never enough. It may be enough to have killed a few minutes, or maybe to have won a free game, but never enough to prove superiority over machine.

Sammy watches their moves, their emotions, then picks one out, someone with a decent score. He seldom picks weaker players, not because he's a great sportsman, but because the "hobby set" just isn't worth the effort; he seldom picks the black leather-jacketed and silver-chained set—no sense causing any more trouble than necessary. Everyone else is fair game.

"Doin' pretty well," Sammy says matter-of-factly, just a trace of admiration in his voice. The player acknowledges the compliment, sometimes only with a shake of his head or an annoyed grunt. "Been playin' long?" asks Sammy who really doesn't care. Soon, the player or Sammy, or maybe someone else, will suggest a friendly game, maybe just a friendly dollar bet. It doesn't make any difference who suggests the game or how much the bet. It doesn't make any difference if Sammy wins or loses; either way, there'll be another game, then another. Sammy lets the mark determine the bet, knowing that greed takes the edge off competence. Sometimes Sammy wins; sometimes he loses. Only he knows when—and why. The number of games won or lost makes no difference; the amount of money won does. It isn't long before he has most of the mark's money, and the mark is accepting a handshake and a free beer.

Others challenge Sammy, thinking the first player or the second or the third just weren't all that good, and Sammy was all that lucky. After all, isn't pinball just a matter of luck? Upon miscalculations like that, dreams are crushed, and Sammy methodically crushes them, taking the bet and stuffing it into a pocket, as though it was no more than a cheap nuisance flyer distributed by some local business. It's chump change, for Sammy will never make his nut by winning five and ten dollar bets. He has a bigger scam, and this one involves the collective greed of just about everyone at the bar. He usually isn't disappointed in mankind's foibles.

From his wallet, Sammy pulls out a crisp hundred dollar bill and politely suggests that his next challenger—whom he had been watching the past half-hour or so—also pulls out a hundred. His challenger doesn't have it? O.K., how 'bout fifty? Thirty? Still too rich? Sammy has a proposal—"You shoot three balls; I'll shoot two and I'll still beat your score." The mark hesitates but takes the bait, and Sammy quietly, professionally, reels him in. The mark wants a chance to get even. Sammy pleads he's running late, doesn't have time for another game, has a real hot date in just a few minutes. But, well—OK, just one more game. It's his last hustle of the evening.

There's money to be made, and Sammy is the undisputed master of the "41 Riff." The mark shoots *four* balls, Sammy will shoot *one* and still beat the score. The mark thinks about it, still has a few bucks left, tries to figure the trick, asks around and no one else can figure the trick, asks Sammy to repeat what he just said, and Sammy again calmly explains what is about to happen. But no one believes him, for everyone thinks he's crazy. Not just a little looped by too much dark beer, but *certifiable*. Ship him to Bellevue. Lock him in a padded cell before he causes himself or anyone else any harm. But first, maybe they should place a bet.

Naturally, Sammy will cover all side bets at 3-to-2 odds; naturally, even the most reluctant see a chance to win a sure fifty percent return in five minutes; naturally, the bartender is soon holding a lot of money. Around the pinball machine, a dozen people are planning their futures on the return from their investment. The challenger—encouraged by friends, buddies, and strangers—

shoots the four balls, and runs up a good score, one that not many people could beat, even if they had five steel balls. But, Sammy only has one, and he must beat that score. He pauses. He sighs. He takes a deep breath. The room gets real quiet. Sammy shoots that last steel ball, and for the next five minutes all it does is bounce around the board, ricocheting from bumper to bumper, thrown back time after time by one of two flippers, constantly in motion, hitting this light then that one, ringing bells and buzzers and who-knows-what-else, sending the electronic scoreboard into overtime as the points stay for the briefest of all possible moments, then fall away for new scores.

Then it's over. The ball finally eludes the flippers and vanishes, but not before the new score is at least a few points more than the old one. The mark stares in amazement. The mark's friends stare in amazement. Just about everyone in the bar stares in amazement. And then they clap and laugh and pat Sammy on the back. They know they've been had. They just don't know how he did it. And Sammy isn't telling.

Only rarely is someone so completely stupid as to want a rematch or a fight. Sometimes, fortified by a half-dozen beers dulling his senses even more, someone will claim that Sammy cheated, but how can you cheat at pinball? Magnets, they guess, but they can't find any, and Sammy is more than willing to play in only a pair of shorts if that's what it takes to root out the skeptics. Most just pay for the lesson. It truly *is* the last game of the evening, for Sammy has absolutely no desire to hang around the bar, not when he's carrying all that cash. He buys a round of beer for everyone, leaves generous tips for the bartender and waitresses, then leaves.

Maybe somewhere someone knows all about him—where he went to school, where he worked, if he's got a family, maybe why he became a hustler, or even if "Sammy" is his real name—but they're not here, not where there are pinball machines and beer. Another bar, another day, another city; the hustle begins again. Like the steel ball, Sammy's always moving, always bouncing from place to place, maybe just waiting until the time he can slide past the last flippers.

I first met Sammy in Newport Beach, California, in 1971. Since then, I have seen him only twice, the last time in 1980.

Three Nights a Week

This is the true story of a better than average musician and a slick out-of-towner entertainer.

It all began one afternoon when the struggling musician lined up a paying audition at a local nightclub. One hour—fifty bucks. If the audience liked what it heard, there would be a contract. Four hours a night; three nights a week; fifty bucks per person per night. Not much, but it was work, and recently the musician's bass was in the pawn shop more than it was on stage.

Hopefully, this gig would last a few weeks, giving him spending money so he could cut back the hours on his day-job as a carpenter. He called up a better than average drummer who was working at a drive-through car wash, and began searching for a lead singer.

Enter, the slick out-of-town entertainer.

At one time, Slick was an opening act lounge entertainer. For years, he had made the rounds of clubs in Reno, Tahoe, and Vegas. But it had been awhile since Slick was on stage, and now he was broke. The musician knew Slick needed money, and Slick knew that the musician needed him in the trio.

Four hours before the audition, Slick showed up for rehearsal, and during the next two hours became very discouraged. The songs weren't right. The arrangements weren't right. The only thing right, of course, was Slick.

Later that evening, the trio went to the club, two of them hoping that Slick wouldn't desert them, and Slick planning to desert faster than crowds leaving a free bar when the booze runs out.

For a fifty-buck hour, the trio gave the audience what it wanted. When the set was over, the club manager agreed to hire the impromptu trio. Four hours a night; three nights a week; fifty bucks per night per person—and if the trio could "bring in a clientele," there would be more nights, more pay.

But it wasn't going to happen.

"I'm not going to do it," Slick told a mutual friend of the bass player and expected the friend to tell the musician. "These guys aren't any good," Slick proclaimed.

"They're tired. They had to work day-jobs. They'll get better. A week of rehearsal and—"

"Place is a dive. Look at the crowd. Bunch of drunks. Tell them I'm out."

"You tell them," said the friend curtly.

The following Thursday night, the musician and the drummer, with a replacement lead singer/guitarist, opened at the club to an appreciative audience. They danced; they drank more beer; they left nice tips. Everything seemed to be going well. But two weeks after that, the club manager said he couldn't afford the trio—which was getting fifty bucks per person per night. "I'm sorry," he apologized. "You guys were good, but I just can't afford it." It was just a matter of finances. It happens all the time.

The next week, the trio was replaced by . . . *Slick* who, the night before the audition almost a month earlier, had left his home phone number with the manager. And, said Slick, he was willing to work three nights a week for a hundred bucks a night—as a solo—thus saving the club about a hundred-fifty a week.

Goodbye Trio; Hello Slick.

In a couple of weeks, many of the faithful customers, the ones who drank buck-a-mug beer and wanted dance music not ballads and folk music, began to leave, and a lot of new customers, who didn't come to dance, began sitting at the tables, buying three-buck mixed drinks and four-buck sandwiches. They didn't care how Slick got his job, or that he wasn't reporting his wages to the union. Slick was happy; the new audience was happy; the club manager was happy, and even raised the wage another seventy-five bucks a week.

But, revenues began falling a little, then a little more. Although the audience was buying more expensive drinks, they were nursing them longer, just sitting and talking and not leaving as much for the waitresses as the beer crowd had. Soon, even this audience slowly began to leave. People can dance to the same music night after night, but they can't listen to the same tunes night after night.

Now, Slick is no fool, and sometime in the past month he had asked another club manager to drop by the club one evening, while there were still a lot of customers. Two months after his first set, after Saturday evening's last set, moments after he packed up his guitar and collected his pay, Slick gave notice. Five days later, he was at a new club. The original club no longer had a Slick, and the revenues fell even further as the new faithful crowd went somewhere else, and the old faithful crowd had long ago deserted, and Slick was somewhere else, making money on an endless journey to capture what once might have been.

MAY 1994

Sounds of the Concrete City

On Fifth Avenue near Rockefeller Center in the Concrete City, a trio of steel drummers beat out a Jamaican tune. The three Jamaicans, in New York because of the "opportunity" they heard exists, arrive every day at mid-morning; they leave every day in the late afternoon, seldom staying until 5 p.m. Very few people stop at 5 p.m.; they have those damned subways and trains to catch, and there's nothing more important than making your connections and leaving the City, even if it's only to Long Island.

Nearby, each with a predefined space against the walls of commerce on this warm Spring day, are a brass quartet, a saxophonist, and a couple of guitar players forced to endure an acoustic lifestyle to avoid getting disturbing the peace citations.

Pleased to receive random applause and the change dropped into open instrument cases, they blend their melodies into the noise of traffic. Maybe a club owner will one day hear them, and sign them to a six-night-a-week contract. Maybe they'll one day play in Lincoln Center or even Carnegie Hall. Maybe a big-shot music producer one day will stop, offer words of encouragement, or hand them a long-term recording contract. Maybe they will be able to leave their crowded Lower East Side or Queens tenement apartments or rowhouse existence and move into one of the outrageous rents of "oh-so-tony" Tribeca or the rarefied atmosphere of the Upper West Side. Maybe . . . There are so many maybes and so few choices.

A few yards away, in counterpoint to the cacophony of the streets, is a flutist, her long black hair tied in the back by a blue and green silk scarf; a peace symbol, conspicuous by its presence in a city that shows little tranquility, caresses her bosom. She sits on the sidewalk, her back against a twenty-one-story spire of steel and glass that blocks out the sun, places an earthenware flute to her lips, and blends haunting Mozartian melodies into an upbeat jazz tempo.

"That's very pretty," I say as she gently lets a coda float off. She looks up, and her eyes smile, telling me, "Thank you."

She continues playing. "You been here long?" I ask.

She stops. Looks at me. Tries to figure me out, then quietly answers. “Since the White Rabbit died.” A strange answer in a strange town. She puts the flute down across an open case that has ten, maybe fifteen dollars, mostly in change. On a good day, street musicians can make more than $100, but this lady probably won’t make half of that today. She looks up, questioning. “Why did you stop?”

“Maybe it was the music.”

“Maybe it was because you’re a reporter.” She has me pegged, but before I can ask her how she knows, she answers. “Your eyes are beautiful, but they’re always searching.”

“Would you mind answering a few questions?” I ask politely.

“No,” she says equally polite, “but I’ll play you some songs.”

“Songs are nice,” I counter, “but I write about people, not songs.”

“Doing a story on the street musicians?” she asks, knowing the answer.

“Maybe. Can I interview you?”

“You don’t need to ask me questions. Listen to the music. It’s all there. There is no secret.”

“Can you at least tell me where you were born? Why you’re here in New York? If you’re a music teacher; member of an orchestra? . . . Waitress? Secretary? Is this a way for you to make some loose change?”

She smiled, a clever somewhat impish smile. “Does it matter? Listen . . . Just listen to the music.” Again she placed the flute to her lips, closed her eyes, and squeezed every emotion from “Where Have All the Flowers Gone?” She looked up. I was still there. Perhaps that surprised her. “Why?” she asked. Just “why?”

“Who knows,” I answered flippantly. “Might be a good story.” She looked at me, searching for her own answers. “I liked the music,” I hastily added. “There’s something more than the music.”

“There’s nothing more than the music,” she said.

“How much can you earn in a day?” I asked. “Do you mind people running past you? Do the police hassle you?” She answered with “The Sounds of Silence.”

Maybe she once had sung of war and protest; maybe she once was a social activist; maybe she still is; maybe she once was an executive, like the thousands who rush past her every day as they try to claim their place in the “me generation”; maybe . . .

All around is the reality of the city, people waking early, grabbing trains and subways, rushing to work, expediting the day, worrying about performance ratings and promotions, lost sales and the stock market, mergers, acquisitions, and even bankruptcies, fortified by a lunch of Maalox and martinis. Perhaps one day, before their first heart attack, they may be able to afford one

of those overpriced $500,000 mid-town condos where they can plant a potted tree on the veranda or move to a home in the suburbs where there are lawns. Perhaps one day they, too, will stop and listen to the sounds of silence.
MAY 1995

Voices of America

In 1974, on the fifth anniversary of the Woodstock Concert, I interviewed several persons who were at Woodstock. During 1970 and 1971, I had interviewed several persons of the Ohio National Guard. Each of the discussions is part of a longer interview. The interesting juxtaposition of their stories—which suggests there wasn't much difference among the American youth, whether soldier or protestor—is a tribute to an age, and to a people.

KENT, Ohio, May 4, 1970—The spirit of last August's Woodstock Nation was murdered early this afternoon in the northern Ohio city of Kent when Ohio National Guard troops opened fire at Kent State University, killing four persons and wounding at least 12 others. The four dead students at Kent State were walking past a 1,000 student demonstration against the war in Vietnam.

"They got their own nation—the Pig Nation. We got ours—the Woodstock Nation. It's where we could liberate our minds from the American bullshit, and just groove on the beautiful vibes. It was our own celebration of life. But the pigs wouldn't let it be. John Sinclair is looking at 10 years for giving two joints to a nark. I don't fuckin' care what their fuckin' law is, they went after him. They went after him 'cause he was leading the Revolution. And they got Abbie Hoffman and Jerry Rubin. Abbie got up on stage and tried to tell everyone all about Sinclair, and fuckin' Peter Townsend tried to throw him off the fuckin' stage. They told everyone to listen to their music, but they didn't have no fuckin' feelings about what was happening."

"I was 20 years old. Same age as them. I didn't get to go to college. Parents couldn't afford it, and I don't get no breaks. I'm working at an auto parts store right now. Makin' a few extra bucks on weekends servin' my country. I put most my Guard pay into a separate account. Maybe some day I'll go to college. Maybe I'll just pack it up and move to California. But, shit, them hippies! They don't know how good they got it. Mommy and Daddy payin' all the bills. Parties all week. They get those high-pay jobs when they graduate. This is the best god-damned place *they'll* ever live in. Ain't no better place anywhere on earth, and all they want to do is tear it down."

"When I heard that the Festival was being moved [from Walkill to White Lake, N.Y.] I was going to pack it up and take a week's vacation. Just leave. I'd heard about all those long-haired hippies and what they do to property. All that dirty language and the nudity, and how they want to destroy our country. But my wife and I, we just couldn't afford to take another vacation, so we took a deep breath and just waited. Y'know what? It wasn't all that bad. Oh, there were a few that thought they were the Hell's Angels, but most were decent and respectable. A lot of us in town made sandwiches and gave water to them. Didn't charge them anything either. A few stores tried to gouge them, though. Charged a buck or two for a 29-cent loaf of bread. Things like that. But most of us just gave them the food. You treat them well, they treat you well. No need to hate anyone. Them and us, we didn't agree on politics. But, now that we seen all that's happened over there [in Vietnam] maybe they were right all along. Maybe they was tryin' to tell us something, and we just didn't listen."

"They called us on a Saturday afternoon [May 1]. I don't think it began as a war protest. I think that they [the students] were drunk and looking for more excitement when the bars closed. They sure as hell made a mess out of that town! Broke damn near every window; set fires in the streets. Don't know why they sent us in. Students do it every Spring. I thought the civil authorities should have handled it, but that wasn't my decision. I just follow orders. Did my job. I don't hate them. Not any more at least. Not after what I seen been happening to us in 'Nam. I had a job to do. Anyhow, the next day [May 2], Nixon says he was sending troops into Cambodia, and that really lit their fuse. Worse than anything Johnson ever did. They couldn't get to Nixon, so they used us."

"I worked in Operations [at the Festival] and I'll tell you this. I didn't hardly crap for a week, it was that bad. I signed up with Woodstock Ventures in May [1969] so I was in on some of the early stuff. We worked 80, 90, maybe a hundred hours a week going into the weekend [August 15-17], and just about around the clock that whole weekend. Man, were there problems! Every kind you could think of. We had a whole fuckin' city to worry about. Medical.

Sewage. Food. Recreation. Hundred departments all working together. Well, mostly working together. John and Joe [John Roberts and Joe Rosenman, festival directors] took a bath on this one. It was really too bad because they really cared, and all they got for it was a whole lotta shit. Do you know that on Friday [the first day] there were roving bands trying to 'liberate' the Festival by breaking down the fences? They could have easily reinforced the fences, but Roberts decided to take them down and make it a free festival. He was afraid people would get hurt. Now, I'll tell you what others would have done. They would have said, 'Fuck it! I'm bailing out.' But these two birds [Roberts and Rosenman], they still manipulated and wheeled and dealed and tried to make it cool. They cared, and they had to eat shit."

"We weren't too organized at first. Best I could recall, none of us knew what was going to happen. All we knew was we were supposed to protect life and property. Lot of us weren't long out of Basic, but we all worked together, kinda looked after each other. Somehow, we all got it set up and organized."

"I remember the Hog Farm the most. They set up these food tents. Ran 'em just like a business. Most efficient operation I ever saw. They were about the most unselfish people you'd ever meet. Hardly none of us had much cash. We just kinda all drifted out when we heard what was going down. The Hog Farm people, they were there cooking and dishing out food for us. And when we'd have some bad trips, they brought us down."

"Bunch of selfish pricks! All they cared about was themselves. Didn't care about no one else. I think about it every moment. Maybe they thought it was *cool* . . . or even *groovy* . . . to shout 'Pigs!' and 'Fuck the establishment!' and go destroying other people's property. They was screamin' and yellin' and actin' just like a bunch of foul-mouthed assholes. Burned the ROTC building, then cut the firehoses. I remember it like it was yesterday. We got hit by rocks and molotov cocktails. Can you fuckin' be*lieve* it! They was yellin' for peace, and they was a-throwin' molotov cocktails! I wanted to take one of their cocktails and give them a fuckin' enema!"

"You get half a million people in 35 acres of open field, and I don't care how good you plan, how many johnnies you set up, it's still going to smell like shit. It was hot and muggy, and that made it even worse. They kept digging ditches, but the odor was still there. Not even all the hash and weed in the world could kill it. I never smelled anything as bad since."

"You know what's funny? With all that yellin' and screamin', and them sayin' how the country sucks, they were still singin' 'This Land is Your Land.' Ain't that a riot? They were singin' patriotic songs and torchin' the flag. Not some Commie flag, but *our* flag. The stars and stripes! Some of the chicks even tried to stuff flowers into our rifles and get us to sing with them. . . . Actually, now that I think about it, it wouldn't have been such a bad idea."

"Man, that music was something else! I mean, like up there in one place was all the best there ever was. The Who. Blood, Sweat, and Tears. Jefferson Airplane. Credence Clearwater Revival. And there was Janis Joplin and Arlo Guthrie and Jimi Hendrix and Johnny Winter and Joe Cocker and Santana and about 30 bitchin' groups, man. And they was a-shakin' that stage, and you could feel the vibes all 'cross the land, and they got into your soul, man—do you know what I mean?—they got into your *soul*—and they stayed there. We were singin' stuff from 'Alice's Restaurant,' and 'Sgt. Pepper.' Whole bunch of us was singin' stuff like 'Hey, Jude.' You know, the part that goes, 'Take a sad song and make it better'? And when we got tired of that, we took off on 'Day Dream Believer.' Just over and over. Mostly just the chorus. But it sounded kinda right. Like it belonged."

"Assholes! They don't belong anywhere. Maybe except on a college campus, protected from reality. Anyhow, they was selfish. Didn't give a shit about nobody. Not their country. Not themselves. No one. . . . Year later, I remember what happened, and it ain't been easy. It hurts. It really hurts, and I get these dreams and I can't make them go away."

"They had this amateur stage away from everything else, and anyone could perform. Bands and poets and jugglers and people who just wanted to have their say. Well, Joan Baez—can you believe it, man, Joan Baez!—well, she sees that there are people kinda just hangin' loose, so she does an hour! A whole fuckin' hour on an amateur stage! Know what else? She didn't just go up on that stage. She waited her turn. Must have waited an hour, two hours. No one knew she was waiting, I guess, but she waited her turn, just like everyone else. Was almost late for the main stage. You know what? Best of them all, the very *best*, that was Country Joe [McDonald]. Now, that's one heavy dude, and that ain't no shit.

Y'know, man, like on Friday, they was a havin' trouble getting all their shit together; afraid there might be some dead air on stage. Well, they find out Country Joe was there, and they asked him to do a set. Now, he ain't on their program, and he was just kinda like hangin' loose. But, y'know what? Country, he just played just about every gig anyone asked him for. Lot of them were for nothing, just to help the cause. Even wrote some good shit, too. Powerful political things. Well, anyhow, last couple of years, hardly anyone seen him. He'd been lying rather low, like on the decline, and he kinda didn't want to do this gig. Well, they told him they needed him. I mean they really needed him. So, he gets out there and follows Richie Havens. Can you believe that? Anyhow, there's a whole shit-load of us out there. More'n Country's ever seen in his whole freakin' life! And he's scared shitless. Man, anyone could see that. And we don't know 'bout Country, him bein' out of circulation so long. But he gave us the Fish Cheer, and we all went wild, just a-yellin' and a-hollerin'. And Country, he was just a whoppin' up there, doin' his thing. I mean, like he couldn't do no wrong. He made it happen. But, wait, that's not all of it. When the big rains came [Sunday], and there were electrical lines all over the place, and all of us were scared shitless 'cause we thought we might fry 'cause we didn't think they'd ever turn off all that electricity fast enough, Country Joe and the Fish get up on the stage and they sang to us. Got our minds off things. None of us could hear him. But he was singing for all he was worth! Calmed us down. That sucker sure as hell knows his shit!"

"My unit had been called up for the race riots in Cleveland [1966] and Akron [1968]. But I'll tell you this, I was more scared at Kent than anytime in my life. They was gonna kill us, like thinking that wiping us out would end the war. We was scared shitless, but we was there 'cause we had to be. They should have let us be. They shouldn't have messed with us. We were dressed in full combat gear. Flak jackets. Helmets. Bayonets. Hell, we never even used bayonets exceptin' for training. And the hippies, they thought they'd take us on. . . . I . . . there's . . . I just . . . can't forget what happened that day. Mom's been tellin' me the past year to just put it all behind me. Get on with my life. But, I can't. I just can't do it."

"When some people think about Woodstock all they think about is the drugs and nudity. The media focused on that. Sure, there were drugs. But we weren't hurting anyone with them. Yeah, I went swimming nude. Lot of us did. Some even walked around nude. It was fucking hot, for Chrissakes! Hot and muggy, and we didn't have showers like the vacationers at Howard Johnson's got. It wasn't no giant sex and drug orgy like the papers tried to make it out to be. It was just a lot of people getting tuned up, just groovin' to the music and free of the Man."

"They just shoulda let us be. You can't commit violence without getting some of it back. I didn't like what happened, but it had to happen. They made it happen. We had to fire! We didn't have no other choice. If we didn't fire, we would have sustained heavy casualties. We told them to disperse. We warned them! We told them what would happen but they kept coming at us, trying to surround us. I couldn't see the looie [lieutenant] and Sarge was in the line somewhere, and none of us knew what the fuck was going down, but we had our orders. They told us to return all fire. Then I heard a shot. Then a lot more. I was sure there was a sniper. I'd have sweared to it."

"They hired hundreds of off-duty cops and guards, told them to protect life and property, but don't hassle us. Kinda just blend in; help out and be cool. Even the on-duty cops from all the local towns, they was cool, rappin' with us. It kinda upset the locals, though, 'cause there's these state laws about drugs, and on the grounds, cops were lookin' the other way, except for the big dealers, and you step off the grounds one foot, and they'd bust you. Locals thought they shoulda been bustin' us no matter where we were, but there weren't enough of them, and they knew we weren't violent."

"They had to be on drugs. All of them. Drugs make you crazy. They were crazy. Destroying everything. Trying to mess with armed troops at bayonet-ready. No doubt about it, they had to be stoned, otherwise why would they destroy everything; why would they try to attack combat-ready soldiers? Don't make no sense."

"I was working in Big Pink. That and the White Tent were the medical problems. The yellow tent was for other stuff. Anytime you get close to a half million people together, you'll have problems. We had to watch for hep [hepatitis] because the sewage facilities weren't much better than open ditches, and the stench—man, the stench was unbearable at times, especially after the rains. All the chemicals in the world didn't get rid of that odor. They don't teach you about this in med school. It's a whole different kind of medicine. Almost no injuries caused by fights. But drugs were all over the place. Mostly acid, hash, pot. Problem was that a lot of the junk was bad. Some of the problem was that a lot of them didn't get too much sleep, not much food. That made it even worse. There were always people waiting for treatment for bad trips or cut feet. Humidity was high. Damned near impossible to work. I guess it was like a combat zone, though I never went. And the casualties were coming in all the time. They had to airlift in supplies, it was that bad."

"No, there was no official order to fire. The situation didn't allow it. A Guardsman always has the right to fire if his life is threatened. It wasn't good what happened, but violence isn't the answer. I . . . I don't regret what we had to do. I was just doing my job. You do what you're told. You can't think about the consequences . . . There was a burst of fire, then it was over. Captain called for a cease-fire almost immediately, but there was still some shooting. Then it was over. 'Ceptin' it's never been over. And it just won't go away."

"Woodstock was a burst of our energy; our music; our spirit. Phil Ochs said it for all of us. He wrote a lot of powerful words. But I cried when I first heard 'I Ain't Marchin' Anymore.' I just sat there and cried. I couldn't take it, it was so powerful, and all I could do was cry. I guess, maybe, that was what Woodstock was all about. It told the world that you could live in peace and harmony and celebrate life and not have to march. And not have to kill. And it's your decision."

"I didn't like what we had to do. There's gotta be something real horrible wrong when we kill our own people. . . . I mean, there we was. Them and us. 'Bout the same age. Next day, it hit me. I just sat down and cried and cried. I couldn't help it. None of us could."

A History Of Fear

Americans have always lived in fear of someone or something. It seems to be a part of our national heritage, like apple pie, baseball, and racism. And now comes Japan-bashing, our nation's obsession in blaming many of our problems on others. It isn't the heinous salaries of American business executives, reduced quality of many of our products, or even the failure of political leaders to lead that's at fault, it's those damn Japanese and all their quality work circles, we argue.

So, we again talk of limiting imports; we assemble around Hondas, surrounded by TV crews, to smash windshields, all of it ably recorded by TV crews; we again talk about limiting Asian enrollment in our graduate schools, obtusely complaining that not only are there too many Asians in our educational system, but they're setting too high a standard for American children to follow. Unfortunately, Japan-bashing is just a minuscule portion of our nation's history of fear. In our country—a leader in research and development, recreation, entertainment, literature, and innumerable other categories—there should be no need to bash anyone, yet it is a part of our mass culture and collective insecurity.

In the beginning, it is written in the grammar school history books, the Colonials came here to escape religious persecution. It was, after all, unfair of England to declare that it had the one true religion when the first Colonials knew they had it all along. Naturally, to preserve peace and freedom, they had no choice but to banish non-believers to the undeveloped colonies of Rhode Island and Maryland.

And, speaking of heretics, when we arrived on this continent, we were afraid that the Indians who preceded us here might actually want to stay on the land they had cultivated. Worse, we were afraid they might even want to co-exist with us. Since we had muskets and they had bows and arrows, we declared ourselves superior—at least we weren't "heathens"—and built forts to make sure the Indians didn't try anything stupid, like move into our neighborhoods.

Then came the French, who hated the British, and soon we could fear Indians *and* French. Naturally, we went to war. When that got to be boring, we became fearful of the British, had a tea party, survived a riot in Boston we called a massacre, and eventually declared our independence.

On our own, we could fear each other, so the Federalists of Washington, Adams, and Hamilton (who now were allied with England) launched campaigns against the Anti-Federalists of Jefferson and his ilk (who now were allied with France). Soon, we were slandering and libeling anything that moved. The result was the alien and sedition laws that were more oppressive than anything the British could have imposed. Speaking of the British, the War of 1812 began two days after England lifted the naval blockade that was a reason why we went to war.

Having once bought Manhattan from the Indians for $24 in trinkets, and a wide chunk of what became our Midwest from France for about $15 million, we now thought Texas might be worth having. By the time that war was over, we owned a large part of the Southwest. And we didn't have to fear Mexicans ever again—unless they, like the Indians and Blacks, were stupid enough to think they might want to live in our neighborhoods.

While many in the North cried for abolition, the reality was that America had divided itself by far more than the slavery question. When it was finally over, there was a permanent blood stain on both our flags.

Then, there were those pesky Indians again who tried to keep us from our Manifest Destiny, our westward expansion. Couldn't they just save us the time and cost of ammunition and kill themselves? Well, at least we put them on reservations so we wouldn't have to fear them actually sitting near us or talking with us. And then we destroyed the reservations.

Of course, there were the Spanish who owned Cuba. When we were through fearing what the Spanish *might* do, we added Cuba, the Philippines, and the right to dig Teddy Roosevelt's ditch across the isthmus of Panama.

And those "uppity colored people"? Wasn't there that Constitutional Amendment? Didn't it release them from slavery? Didn't we even give them "separate but equal" status so they wouldn't be tempted to become doctors and lawyers and teachers—and reporters—and possibly drink from our drinking fountains, eat in our restaurants, move in next to us and marry our sisters?

With the massive increase of immigrants, we were afraid of losing our jobs to "foreigners"—or worse, having to (yeah) live with them. They were, after all, *different*. Those Irish fleeing famine. The Italians and eastern Europeans looking for a better life. The Jews fleeing political and religious oppression. We had to make sure they knew their place, 'lest one of them thought that becoming a physician in America might be a better calling than selling vegetables off a cart in New York City. So, we imposed quotas on college admissions, and carefully explained that Irish had to be cops, eastern Europeans had to be coal miners, and Jews were supposed to be tailors and entertainers.

We feared the Germans so much in World War I, a not unreasonable fear, that we wouldn't even eat sauerkraut and frankfurters—it had to be Victory Cabbage and hot dogs. We feared the Germans again in World War II—hey, maybe they're on our side after all; aren't they killing off the Jews?—and added the Japanese and Italians. Into concentration camps we euphemistically called relocation camps, we resettled Americans of Asian ancestry. Into "separate but equal" military units, we put the people we still called colored soldiers while all of us proclaimed we were fighting for democracy.

After the war, we feared the Cold War and godless Communism—ever notice how we always tied "godless" and "Communism" together?—and in our fear of the unknown, like we had during the witch trials, went searching for Communists under our beds. During the next few decades, we increased our fears. We feared the Vietnamese. The Arabs. The possibility that we wouldn't get enough oil, and Exxon would only get four billion dollars of profit a year instead of five billion.

While proclaiming our individuality, we try our best to look, act, and think like everyone else, lest someone label us "different" or, worse, "radical." We fear ideas that aren't what we believe, so we continue to ban and burn books, whine about the National Endowment for the Arts, and protest Murphy Brown's single motherhood lifestyle, forgetting that our nation was founded upon a Libertarian principle that *all* views should be heard.

In a nation that seems to value appearance over intellect, a nation where there are no ugly anchors on TV, we are so afraid of not looking at least as well as anyone else that we spend billions for makeup to cover blemishes; spas, gyms, steroids, and scam diets to "tone up our flab" so no one scorns us for being fat; and billions more for augmentations to fill out, liposuctions to reduce, and Preparation H to shrink our wrinkles.

We don't hire the handicapped, the short, the tall, the fat, the skinny because they're "different." We fear gays, who we call "fags" and "queers," because they're different—or, we hope, different from the rest of us.

We are so afraid of not being "cool" that we allow advertising to dictate what we wear, what we eat and drink, and even what we drive. We are so afraid that someone else will get something more than we have, we buy larger stereo systems, more expensive cars, and bigger houses.

We go to college because we're afraid we won't get a good job, then spend 40 years on that job afraid to do anything different, creative, afraid to speak out for fear of displeasing someone who might fire us.

We buy .357 Magnums so we can blow away burglars—or in fear of neighbors who take short-cuts through our back yards at night. Or to murder people whose views are different from ours.

We fear workers getting a piece of the pie, our pie, so we swallow it whole to make sure no one else can get any. We attack teachers because they get the Summer off and we have to work. We attack the news media because they don't report what we want to hear—or believe. We attack the unions because we're afraid they're getting more for their members than we're getting for ourselves; they're "greedy," we sob when we should be attacking ourselves for not fighting for equality for all of us in the workplace.

And now comes the Japanese once again, with their borrowed electronics technologies, downsized cars, and quality control workplaces. So, we're going to limit imports, complain that there are too many Asians in graduate schools in America, and again put up signs that read "No Japanese Need Apply."

Franklin Delano Roosevelt said that the only thing we have to fear is fear itself. Perhaps, it's now time to take responsibility for ourselves and say that the only thing we need to fear is ourselves.
AUGUST 1993

'And Now a Word From Your Local Sponsor'

In the *Police Academy* movies, Michael Winslow portrays a cop who baffles and harasses the bad guys by imitating the sounds of helicopters, machine guns, and even the high-pitched sirens of police cars. Naturally, the sixty trillion or so people who have seen the movies know *Police Academy* is a comedic exaggeration. Unfortunately, the nation's military and paramilitary agencies haven't yet figured this out.

In 1990, the Army set up loudspeakers to blast rock music at opera-loving drug-dealing Panamanian dictator Manuel Noriega, figuring that no one could withstand "Voodoo Child" and "You're No Good" for long. During the Gulf War, the army set up speakers to transmit sounds of tanks to make Sadaam Hussein believe an invasion was imminent.

And now comes the Bureau of Alcohol, Tobacco, and Firearms (BATF) and the FBI which also must think that *Police Academy* is real. Having bungled their way into a month-long stand-off— having decided that massive paramilitary force was preferable to a subpoena which could easily have been served on the cult—they're now trying to oust Wakko David Koresh and his Branch Davidians from their Waco, Texas, fortress by using psychological warfare. It may be the only weapon left since it seems as if the hundred cult members still in the fortress have better position and more weapons than the FBI, Army, National Guard, several Sheriff's and police departments, and the BATF. Outside the compound, the feds have shined spotlights into the compound, and set up a ring of loudspeakers. They recorded a high-pitched screech from an off-the-hook telephone, cranked up the decibels, and beamed that noise into the compound. They have also blasted Christmas carols, chants from Tibetan monks, military bugle calls, and rock music at the man who believes he's a prophet, a savior or, depending upon the tide, the son of God. The spotlights and rock music haven't seemed to affect the Davidians. Since the Feds cut off the compound's electricity, the cult members say they welcome the evening light. And, since they have an arsenal, they also have mounds of earplugs which they have been using to block unwanted sounds.

So far, it seems the only ones affected by the psychological warfare are the feds, a third of whom may now be in rest homes. Nevertheless, should there be any more militant religious fanatics out there, the government should know there are far more effective methods it could use to force the cult members to surrender.

The feds could send an insurance salesman to the front door. "Hi, have you given any thought to where you're going to spend eternity, and by the way, does your family have enough money to bury you?" Maybe a used car salesman. "Hey, David, baby, got just what you're looking for. Got a great buy on an '85 M-13 Armored Personnel Carrier. Great condition. Fantabulously low price. Used only a couple of times by a little old FBI agent who drove it only to an impending apocalypse."

Perhaps we can convince him he's getting a Lifetime Achievement Award from the National Rifle and Howitzer Association. When he emerges from the compound to collect the bronzed AK-47, the feds nab him.

But, for true and unlimited torture, the kind that Amnesty International would surely oppose, I'd set up large-screen TV sets all around the fortress. I believe I have the ultimate "surrender-at-any-price" programming.

I'd open the broadcast day with a sermonette by personality-drenched FBI director William Sessions, followed by a 15-minute quick-cut montage of politicians and government officials saying, "Trust me." Next would be an hour of clips from sixty or so of the 15,000 celebrity exercise videos currently on the market, followed by an hour of clips of "Stupid Funny Home Video Tricks," and a couple of hours of talk shows, featuring men who think they're gods and the women who think they're virgins.

Next would be a rerun of "Cop Rock," followed by music videos of Ratt, Nine-Inch Nails, and Ice-T. Inbetween the programming, we'd beam endless commercials of Sally Struthers blathering about correspondence courses, some bald-headed guy telling how to restore hair, and Ed McMahon telling every known life form that it *could* be a winner.

About 6 p.m., bring on the local "Happy-Time News," with Susie Sweetwater's big scoop on whom she thinks will win the next

Miss America contest, Harry Handsome's biting feature about the county's largest cucumber, and Darla Dormant's questions to Waco residents—"How do you feel about having a lunatic religious fanatic murderer in your town?" If that doesn't cause Koresh to surrender, unleash the "Big One," guaranteed to bring him to his knees to pray for eternal salvation from TV—back-to-back reruns of the three Amy Fisher movies.
MARCH 1993

On April 19, 1993, the FBI and BATF stormed the compound in Waco. Seventy-two persons were killed, most by suicide or by other members of the compound, some by the government. In a subsequent investigation, both BATF and the FBI came under extensive scrutiny for their roles in applying excessive force to a situation that might have been resolved peacefully.

War in the Gulf: Lessons of an Obsession

With flags flying, Buck Rogers technology, and an allied army of 500,000 from more than two dozen countries, led by a charismatic teddy bear general, we sounded forth the trumpets, declared we believed in democracy, and systematically decimated Iraqi forces in a one-month air and artillery barrage, then Hail Mary-ed them in a four day ground assault in February 1991.

When it was over, and 25,000-50,000 people lay dead, and another 150,000-200,000 injured, thousands of them critically, we proudly declared we defeated not only the "Butcher of Baghdad," but dictatorships and territorial aggression as well. So, let's see what we learned and what has happened since the United Nations Memorandum of Understanding of April 18, 1991, ended the Persian Gulf War.

Shortly after the Iraqi invasion the previous August, many Kuwaitis, including most of the royal family, fled to Cairo, London, and other capital cities where they took their wealth, complained about Iraqi atrocities, partied, and waited for the Coalition to hand them back their country. There is little question that Iraq committed atrocities, not only against its own minorities but also against Kuwaitis. But, the Kuwaitis aren't exactly saints. A Human Rights Report issued by the U.S. Department of State at the end of 1993 pointed out there are "continuing reports of torture and of arbitrary arrest, as well as limitations on the freedoms of assembly and association."

Four years before the Iraqi invasion, Kuwait's royal family dissolved Parliament. Under American pressure, the al-Sabah royal family, safely in exile and promising anything to the Coalition, agreed that once the war was over it would permit elections. In October 1992, the amir finally restored a 50-man parliament, but political parties are still banned. The "democracy" we fought to preserve is for the 90,000 or so voting citizens of Kuwait. More than 550,000 Kuwaitis—including women, servants, and laborers—have no voting rights. Only males older than 21 who can trace their ancestors to having lived in Kuwait prior to 1920 may vote.

The Kuwaiti Constitution, suspended in 1962, still hasn't been reinstated. Freedom of the press was curtailed in 1985, then barely restored after the war. Although prior restraint upon the media has been abandoned, the media and people still face severe restrictions. A Department of Censorship exists. The people and the media may not criticize the amir, the Kuwaiti government, or the Islamic religion.

From the war, we also learned that our military intelligence wasn't as precise as we needed. While the media parroted almost daily official statements that precision bombing by the Coalition had destroyed all of Iraq's SCUD missile bases, the SCUDS somehow kept popping up, often against Israeli targets. We also learned that the highly-touted U.S.-built Patriot anti-missile missiles were about one-third effective.

We reported we destroyed the Iraqi military, then were surprised when Saddam, apparently with little opposition, launched attacks upon Shi'ite and Kurdish populations within Iraqi borders.

The military told us, and the media reported, that Iraq had, and was likely to use, chemical, biological, and possibly nuclear warfare. In contrast, the American military piously proclaimed that we, noble warriors all, would never use those weapons. But, the media either didn't know, or they didn't report, that during the war, the Coalition had limited nuclear weapons in Saudi Arabia, driven as much as 500 miles from the ports to the front by American soldiers and Marines.

We learned that the military bunker in Baghdad we bombed really did hold hundreds of frightened women and children. Our military and press claimed Saddam had deliberately put them there as hostages. Saddam declared, with some justification, that they were in a heavily defended bunker to be safe from American attacks. The baby milk plant we destroyed, then said was a front for weapons manufacturing, really produced baby milk.

In a month of saturation bombing, Americans were subjected to saturation media coverage. We learned that although many reporters excelled under extreme difficulty, most of the 1,800 onsite reporters didn't have a clue of how to cover a war, asked dumber questions than freshmen after an all-night beer party, and allowed themselves to be manipulated by one of the most efficient PR operations ever devised in American history.

Seventeen reporters went to Saudi Arabia in August 1990 at the beginning of Operation Desert Shield. By the end of the month-long Operation Desert Storm at the beginning of the next year, almost 2,000 reporters were crawling all over themselves to be first on the air or in print with the "latest" in war news. The reality was that CNN usually had the latest and most comprehensive coverage. Fortunately, this was a short, easily-managed war, unlike the seven years the U.S. was bogged down in Vietnam's rice paddies and jungles.

In the Vietnam War, the American government realized, perhaps too late, that it needed to win the "hearts and minds" of the enemy. In the Gulf War, the American government realized it had to win the "hearts and minds" of Americans, especially since this war, unlike any other war, could be reported live anywhere in the world.

The Pentagon quickly and efficiently established two Joint Information Bureaus in Saudi Arabia, and set up two-a-day briefings, to provide as much information as most reporters could handle—and, possibly, to make sure that the reporters would be so bogged down in certain information that they wouldn't have time to find "other" information.

The Pentagon didn't want pictures of body bags being prepared at Dover Air Force Base, Delaware; there weren't pictures of body bags. The Pentagon didn't want reporters to talk with the military, private through general, or Arab civilians unless an "escort officer" was present; the media didn't do 1-on-1.

To assure the media didn't overstep its boundaries and jeopardize the security of the troops, the Pentagon established "media pools," the selection of a few reporters, mostly from the "major" organizations, to accompany the troops and to provide coverage that all media could take and disseminate. "Pools rub reporters the wrong way, but there is simply no way for us to open up a rapidly moving front to reporters who roam the battlefield," said Pete Williams, assistant secretary of defense for public affairs. He claimed the use of the pools got "reporters out to see the action . . . guarantees that Americans at home get reports from the scene of the action, [and] allows the military to accommodate a reasonable number of journalists without overwhelming the units that are fighting the enemy." It also assures that there would be fewer stories.

Nevertheless, said Williams, "Our goal is to get as much information as possible to the American people about their military without jeopardizing the lives of the troops or the success of the operation." It was a decidedly different military philosophy than the one in Grenada which Williams readily acknowledged as "a journalistic disaster." As a result of complaints against the military's refusal to allow media coverage of its mini-invasion of a Caribbean island, apparently to make the world safe for American medical students, the Pentagon created a quick-strike media pool which, said Williams, "could be called upon on short notice to cover the early stages of military missions."

Finally, as much as we learned from this war, we learned a lot about propaganda and public opinion. In October 1990, two months after Iraqi troops invaded Kuwait, a 15-year-old girl testified before a Congressional committee that she was a hospital volunteer who saw Iraqi soldiers enter a Kuwaiti hospital and "take the babies out of the incubators, and left the children to die on the cold floor." Several members of Congress said her testimony helped them decide to support what became Operation Desert Storm less than three months later.

Within months of the end of the war, we learned that the 15-year-old wasn't a volunteer but really the daughter of the Kuwaiti ambassador to the United States, and that there may not have been an atrocity at all. We learned that her testimony was orchestrated by the New York PR firm of Hill & Knowlton which received a large chunk of more than $20 million that Kuwait paid to establish two American-led front groups whose missions were to convince the American government—and, of course, American business—that there was overwhelming support by the American people for war against Iraq to preserve what we thought was "democracy," but was probably America's obsession with oil for energy.

For that, we sent in hundreds of thousands of Americans to kill, and to be killed, by hundreds of thousands of Iraqis, and to pretend we were fighting to preserve a democracy that never did and probably never will exist.

MAY 1994

A Medal for the Army

More than four million medals were awarded to Americans who participated in the Gulf War. A high-ranking Army officer, trying to justify the cost for what might have been a bad case of medal-inflation, said that the awarding of the medals was not only good for morale, but also good public relations for an Army that had severe morale problems following the Viet Nam War. For most of the military in the Gulf, it was not only their first taste of combat, but also their first medals as well.

Once soldiers lose their combat virginity and realize that promotion is based upon how colorful their chests are, they're going to want to continue to invade countries and humiliate peacocks. The Pentagon is only happy to oblige, as this semi-official not-so-secret transcript of a recent Pentagon meeting reveals.

"Col. Klunk, as our top PR advisor, I believe you have some ideas on how to get more medals."

"Thank you, Gen. Kuhster. The Gulf was a PR coup for us, so we need to intensify our existing positive behaviors by reversing counterindicative negative behavioral attitudinal objectives."

"In English, Colonel."

"We follow it up with the same thing. After all, in PR, imitation is the sincerest form of guaranteeing you won't have to have an original idea."

"You want us to invade Iraq again?"

"No, sir! In PR, duplication is *bad*; *imitation* is good. We invade Saudi Arabia. Same weapons and tactics. Different country. Wipe them sheiks clear into democracy and force them to grant equality to *all* their citizens!"

"But, Colonel," said the chairman of the Joint Chiefs of Staff, "if Saudi Arabia becomes a true democracy, it would have to allow women and gays to serve in combat units."

Col. Klunk immediately saw the problems of a democratic invasion, but was undeterred. "Bosnia and Rwanda! It's a two-fer! Why, them poor SOBs facing ethnic annihilation need all the help they can get. We'll call it a global *humanitarian* mission. Those weeping sister knee-jerk liberals on the eastern papers would fall for that in a drop of a politician's morals."

"Not good," said the general. "Our ambassador from Exxon doesn't think Bosnia or Rwanda have any resources worth protecting."

"We'll find another country in Africa. Makes no difference which one. None of the media give much coverage to Africa anyhow. We could invade the whole continent, and the media would still be covering O.J.'s trial."

"Africa's out," said Gen Kuhster, reminding Col. Klunk that even the Army needs probable cause to launch an invasion.

"Haiti."

"Two problems. First, most of the Haitians are in Florida or on boats somewhere. Second, even with the world's best super computers, none of our staff can move as quickly as the Administration's policies change."

"Let's run this one up the flag pole and see who salutes it," said the military flack. "We launch a Defcon 1 invasion of Colombia, neutralize the opium dens, and make the world safe for lite beer. More than anyone else, the media will love us for that!"

"Can't do that either, Colonel. Half the White House staff would turn their noses up at even the slightest suggestion that we curtail the coke trade."

"We invade France. I never did like those snooty snail-eaters." Gen. Kuhster shook his head, explaining that no matter how arrogant the French are, they're still allies.

"Some allies!" whined Col. Klunk, "They don't even let us invade their country so we can boost morale."

"Now, Colonel," said the chairman soothingly, "the Pentagon isn't opposed to sending combat troops all over the world, we just don't like the countries you've suggested."

"I know where you're coming from," said the flack, "and I appreciate your input into this, but let me share this possibility with you. We develop a series of attitudinal objectives, take a proactive stance, and target our selected publics."

"In English!"

"We find out what people Americans don't mind wiping off the face of the earth, then invade *them*!"

"You may be right," said the chairman of the Joint Chiefs. "We need to find the most worthless piece of real estate on earth where nothing useful happens, where double-talking greedy little toads are running the show and are too tied down by scandal to even mount an opposition. We do that, and we'll have good PR—and more medals."

And that's why, on a humid Summer morning, a crack team of Navy SEALS, a Ranger company, Delta Force, and two divisions of combat-trained ribbon-bedecked soldiers were seen marching down Pennsylvania Avenue on their way to the Capitol.

JULY 1994

Chicken Advertising

A recent study indicated that the U.S. Army spent about $100 million last year for advertising, about twice what Kentucky Fried Chicken spent. No matter how it's figured, the Chickens get better cluck for their buck.

And so one day the Secretary of the Army, several of his top aides, a few assorted generals, a marketing specialist, a communication consultant, and a gold-chained advertising account executive met in a top secret session.

The marketing specialist, with 20 feet of computer paper spread across two tables, babbled endlessly about chi squares, gamma coefficients, multivariant analyses, and focus groups. He thanked the generals for their time, dropped a four-figure bill in their laps, then abruptly left for another meeting, probably with the Navy to keep them from falling further behind the Marineland budget.

The communication consultant suggested that sergeants should do a better job of "bonding" with their recruits. Moments later, he was recalled to active duty and sent to Alaska. It was now the adman's turn.

"Larger billboards?" the adman weakly suggested, hoping not to be sent to his former assignment in Yugoslavia where he handled both the Bosnian and Serb accounts.

"What do the Chicken people have that we don't have?" asked the Secretary.

"Their Colonel?" he lamely suggested.

"Their *Colonel*?" thundered the Secretary. "We have colonels of our own. And generals! All those generals running around here are a menace!"

"I don't think you understand," said the adman. "Their colonel is a nice grandfatherly type that people can relate to."

"What about our *uncle*?" scowled the Secretary.

"He was a real cool dude when we had the draft," said the adman, "but now that we have a volunteer army, I don't think Uncle Sam is the appropriate image. After all, he's got that beard and funky hat and— "

"Do we have any admen in Iraq, yet?" the Secretary asked an aide.

"No problem," said the adman wiping his brow. "How about increasing the choices a soldier has?" After the laughter died, the Secretary explained that soldiers don't have choices. "Chickens have choices," said the adman. "There's also regular and crispy."

"Think!" the Secretary commanded. It wasn't something the adman planned to do, but if it would keep him from exchanging silk Gucci shorts for olive drab, he'd try thinking.

"Family plans?" the adman meekly suggested. When no one said anything, he brazenly continued. "We develop a *real* family plan. Mother. Father. Two kids. All in the Army. Remember, the family that shoots together is the family that dies for their country together."

"I'm not sure it'll fly," said the Secretary. "After all, Congress is trying to cut us back."

"It flies for the Chickens," said the adman. "Their family buckets are very popular."

"If we wanted to be popular, we'd have kept the communications guy," said the Secretary. "We're spending too much and getting too little. We're in a slump."

"When the Chicken has a slump," said the adman, "it has a sale. How about a 2-for-1 sale? Maybe two guns for the price of one. Two pairs of combat boots or whatever."

"Sometimes *one* adman is twice what we need," moaned the Secretary who began signing travel orders.

"No biggie," said the adman. "I'll get a line on something. Lines! That's it! Soldiers are always complaining about lines. We claim that the Army doesn't have lines. No lines. Faster service."

"I'm sorry," said the Secretary, closing his note pad, "but lines are what makes this Army great. Fortunately, with the budget cuts there will be even longer lines."

"Sir," said the adman ominously, "don't you want parity with the Chickens? It could force Congress to stop your cuts and increase your budget."

"Parity" was apparently the magic word. The Secretary stood up, looked at all his aides and generals, and boldly announced that the misnamed Army Intelligence Corps will immediately begin surveillance on the Chicken Place. "And as far as this advertising account executive is concerned," said the Secretary, "give him the best suite in the Pentagon. He could very well be our secret weapon in this war of image."

APRIL 1993

Kernels of Truth

It was an important news day. The massacre in Rwanda, which followed a period of stability following a three-year civil war, continued with no hope for peace, and would claim 500,000 more lives. In Somalia, warlords were still fighting small battles for political control. In Haiti, soldiers murdered two dozen fishermen. Serbian artillery was still hammering targets in Bosnia. South Africa politicians were campaigning for office in the country's first election in which all citizens could vote. And, in America, thousands still weren't receiving adequate health care, President Clinton called for Social Security to be taken out of the department of Health and Human Services and become an independent agency, and Americans were planning to go to Yorba Linda, Calif., to pay tribute to former president Nixon who had died only a few days earlier.

But, the most important news story, the one CNN and the networks pumped all day, the one that was splashed in inch-high types across six columns on newspaper front pages, told a shocked America that theater popcorn was dangerous.

The Center for Science in the Public Interest told reporters that a medium bag of popcorn, popped in coconut butter, had more fat than a day's menu of a bacon-and-eggs breakfast, a Big Mac and large fries lunch, and a steak dinner with baked potato and a mound of sour cream. Add a few wisps of butter, and you need a Roto-rooter to clean your arteries. The Council claimed that theater owners use coconut oil because it gives popcorn a unique theatrical flavor that the almost-healthy corn oil or air popping can't.

After learning that cigarette smoke is dangerous, thousands of Americans stopped smoking after sex. But, because of AIDS and other venereal diseases, they also decided to abstain entirely. So, they went to R-rated movies, but now we learn they can't even munch popcorn, and must now practice Safe Kernel.

Because about 90 percent of all movie house revenue is from food sales, the impact could force theaters out of business. If attendance at movies drops because of a drop in popcorn sales, then thousands of actors, costumers, and film projectionists will be thrown into the unemployment lines. There's nothing more pathetic than seeing Clint Eastwood whimpering to an unemployment clerk about how hard he tried that week to get a decent minimum-wage job that would make his day.

Those addicted to popcorn might be able to smuggle in a bag or two. Naturally, theaters, to preserve their dwindling profits, would have to hire popcorn police to make sure that didn't happen. Since there's no ban on assault weapons, the Kernel Kops could spray a few thousand rounds of 9 mm. cholesterol pellets against popcorn sneakers.

Some governmental body somewhere will soon prescribe there must be separate sections of the theater for those who eat popcorn and those who don't. After all, secondary popcorn smell can be almost as devastating as drinking a gallon of coconut oil. At the very least, there'd have to be warnings on popcorn bags, like, "Eating too much popcorn can make you go blind" or "Making noise while eating popcorn can injure your eardrums and annoy that spike-helmeted biker nearby."

Based upon the evidence, Congress will convene a special select subcommittee to analyze the impact of popcorn upon society. The first witnesses, of course, will have to be Orville Redenbacher and that Howdy Doody grandson of his. Certainly, in our gold-chained society, anyone who wears bowties must be suspect. If the Cold War still existed, the Popcorn Potentate would probably have been a Communist. After all, doesn't the name "*red*enbacher" really mean "Commie backer"?

Recently, a cigarette company executive, with straight face, told a Congressional Committee that cigarette smoking is no worse than eating Twinkies. If he had said that cigarette smoking wasn't as bad as munching theater popcorn, he'd probably be a hero of the industry by now.

Of course, producers will continue to churn out R-rated movies with 62,000 violent acts per hour, and feel rather safe about their Constitutional license to practice freedom of celluloid.

APRIL 1994

Tanness, Anyone?

Between a carpet of knee-deep snow and a nimbostratus ceiling, one of my friends is still sporting her Summer tan. I know it's phony—and she knows that *I* know it's phony—but I have long ago stopped teasing her about it. In her never-ending quest to appear to be beautiful and healthy, she has slathered skin tanning lotion into every pore of her body, laid out on roofs and beaches to catch whatever ray was passing by, and goes to a tanning salon twice a week. I'm not sure she's ever stepped into the surf.

For decades, I have endured the scorn of these fake-skin friends, their hair bleached to fireball yellow, their skin tanned to the color and consistency of obsidian, as they sweat their lives away, ruled by the nation's mass media which give them myriad examples of what editors think are the "beautiful" people. If the editors would just get skin cancer and shut up, the world might be a better place. Nevertheless, I have always been content to know I had more genetic pigment than all of them, and don't need to cremate myself on a rooftop to be healthy. A "natural 7," I reason, is far better than crapping out with a "phony 10."

Once, women desperately wanted to look pale. Ashen was to be admired. Pallid was wonderful! The lighter the skin, the healthier they believed they were, even if it meant hiding in a basement and fighting any attempt by Vitamin C to force its way into their lives. These women would read *Macbeth* and admire the ghost. Any darkness of the skin reflected that they weren't women of leisure, but (*horrors*!) *working* women—the kind who go out of doors and have to (*shudder*!) *do* things.

Then, it changed. Some universal force, probably sitting under a sunlamp in a New York publishing house, now decreed that women of leisure got suntanned. Not just a little suntanned, but an I'm-darker-than-Bill-Cosby-tan. Ten minutes, an hour—what the heck, make it an entire afternoon! Soak up those ultraviolets! And when they couldn't find enough sun to char their skin and fry their brains, they bought sunlamps, reflectors, and gallons of lotion, guaranteed to make their friends believe they had just returned from a decade in Bermuda—or Nigeria.

And then came tanning salons. In the semi-privacy of a casket, people could pay twenty bucks for fifteen minutes, slobber even more lotion on themselves, and look even healthier! Have you ever seen what a couple of hours a day in the sun can do to an unprotected body over a few years? If you don't have to chase away knife-wielding scouts from the Tandy Leather Co. from trying to skin you, then you have a chance to live until a ripe old age of at least 40. And if Tandy doesn't get you, there's a pile of myeloma waiting. Ever see what cancer of the eye or ear looks like? Ever see a jellyfish on a rotting log?

Cancer scare? There's still sunblock. Just pick a number. Any low number. You'll "protect" yourself and darken up just like that Ban de Soleil model—and look just as good. After all, would advertising agencies lie?

Now, while many people desperately want to have dark skin, they aren't willing to appear to be "ethnic." So, just in case someone could confuse them with being Black, Hispanic, Jewish, or any other genetically dark-skinned type, they dye their hair blonde. Just as they believe that the advertising agencies wouldn't deceive them, they believe that blondes have more fun. Didn't that great American philosopher Lady Clairol say it? It must be so. And, of course, there are about 65,000 solutions on the market, each advertised as the greatest miracle since coconut oil, just designed to make you have fun while you lose every follicle in your genetic pattern.

OCTOBER 1993

Cramping Up for Health

Athletes and nerds do it.

Teachers and pipe fitters do it.

Topless waitresses and ministers do it.

Even Bill Clinton does it.

On bicycle paths and highways, on dirt roads and in parks, in heat, rain, and snow, people are darting in front of cars, chased by dogs and preppy muggers, in a never-ending quest for a faddish youth and the right to believe they match up to all the media models and their exercise videos.

And so it happened that one cool morning while walking in the park I was caught in the middle of the morning madness. From out of nowhere hundreds ran past, sweating and straining, air cushions on their soles, earphones on their heads.

I was spun around several times. Dizzy, I tried to regain my composure, only to be hit again, as if the runners were oblivious to everything else. At first frightened, but now angry, I tried to get up, only to be knocked down by another cluster of runners. Finally, I trapped one. Actually, he tripped over me.

"Confess!" I shouted. Why do you run? What compels you?"

"Health," he said, panting and wheezing. "Makes your heart beat faster. By beating faster, it exercises it. Becomes stronger. You get a resting rate of 50. Maybe 60."

"Why couldn't you just take a pill for it? Lots of pills will make your heart beat all over the place, then stop on command."

"Must experience pain," he wheezed.

"Do you experience a lot of pain?" I asked.

He took a series of small breaths, then told me that his feet hurt a lot more since he started running. He even blurted out that so did his ankles, thighs, knees, arms, neck, and even stomach. "But we all have to go through it to make us healthy," he said. "By staying healthy, I can work longer hours so I can buy more things."

"Like what?" I asked.

"Like these track shoes," he said, pulling one off a swollen foot while massaging a muscle cramp. "These shoes are made specifically for runners. They're completely cushioned, with special arch supports, a reinforced toe, and special color stripes to give it that extra special look of authenticity. They only cost $159.95, too. A real bargain."

"And they last for years," I said admiring the canvas.

"Oh, no! They last three, maybe four months. Even with these custom-made ten buck running sweat socks, all that sweat can really wear out a pair of shoes."

"I notice that you're wearing a floppy hat. Is that a runner's hat?"

"Sure is. Cost only $9.95, but worth every penny."

"The headband and wristbands?"

"Ten bucks."

"The wrist wallet?"

"Only fifteen. Twelve and a half when they're on sale."

"The suit?"

"If we don't have the proper suit, we'd be humiliated in front of our friends and neighbors. This one cost only $129.95, but it'll last at least another five, maybe six months. Besides, these are specially made to allow us to perspire more. That way we lose more water. Five, ten pounds at a time."

"That's a lot to lose at one time, isn't it?"

"If it wasn't for the salt pills, we'd lose even more."

"How much?"

"A buck a pill. You need two, three pills a day. Then, there's the mandatory stop at Duffy's."

"The tavern on Hydroponic Avenue?" I asked.

"Sure. You don't want to lose too much fluid. That isn't healthy. So, after we run, we all stop over at Duffy's to get a few drinks and buy some more pills. Nice fellow, that Duffy."

"You drink *beer* after a healthy run?" I asked suspiciously.

"Beer. Whiskey sours. Carrot juice. Whatever keeps the machine going." He paused a moment, his breathing down to only 30 puffs a minute, then continued. "I'd sure like to talk with you longer, but I don't want to be too far behind. It's bad for the image. Besides, I've got to get to work. Make more money. Buy more running clothes."

"What do you do?!" I shouted after him.

"Advertising executive," he shouted back. "If you get a chance, stop by Duffy's for a drink. He'll also give you a good buy on track shoes, sweat socks, hats, sweatbands, wrist wallets, and running suits."

NOVEMBER 1992

Death by Healthy Doses

They buried Bouldergrass today. The cause of death was listed as "health."

Bouldergrass had begun his health crusade more than a decade ago when he moved from smog-bound Los Angeles to a rural community in scenic green Pennsylvania, gave up alcohol and a two-pack-a-day cigarette habit, and was immediately hospitalized for having too much oxygen in his body.

To burn off some of that oxygen, he joined two-thirds of America's "beautiful people" on the jogging paths where he believed he was sweating out all the bad karma. In less than a year, the karma left his body which was now coexisting with leg cramps, fallen arches, and several compressed disks. But at least he was healthy.

To make sure he didn't get skin cancer from being in the sun too much, he slathered four pounds of No. 35 sunblock on his body every time he ran, and went to suntan parlors only twice a week to get that "healthy glow." He stopped blocking when he learned that suntan parlors weren't good for your health, and that the ingredients in the lotions could cause cancer. So, he wore a jogging suit that covered more skin than an Arab woman's black chador with veil—and developed a severe case of heat exhaustion.

To further his quest for health, he began lifting weights and playing racquetball six hours a day. Four groin pulls and seven back injuries later, he had just two percent body fat, and a revolving charge account at the office of his local orthopedist.

His dietary habits were equally exemplary. Several years earlier, Bouldergrass had stopped eating veal as part of a protest of America's inhumane treatment of animals destined for supermarkets. Now, in an "enlightened" age of health, he gave up all meat, not because of mankind's cruelty to animals, but because a TV panel revealed that vascular surgeons owned stock in meat packing companies. Besides, it was the "healthy" thing to do.

For a couple of years, lured by a multi-million dollar ad campaign and innumerable articles in the supermarket tabloids, Bouldergrass ate only oat bran muffins for breakfast and a diet of beta carotenes for lunch, until he found himself spending more time in the bathroom than at work. He eliminated the muffins entirely after reading an article that told him oatmeal, bran, and hood ornaments from Buick Roadsters were bad for your health.

Bouldergrass gave up milk when he learned that acid rain fell onto pastures and was eaten by cows. When he learned that industrial conglomerates had dumped everything from drinking water to radioactive waste into streams and rivers, he stopped eating fish. Then, he gave up pasta after reading about all the creepy crawlers who become fat from dough.

At the movies, he smuggled in packets of oleo to squeeze onto plain popcorn, until he was bombarded by news stories that revealed oleo was as bad as butter, and most theatrical popcorn was worse than an all-day diet of sirloin.

When he learned that coffee and chocolate were unhealthy, he gave up an addiction to getting high from caffeine and sugar, and was now forced to work 12-hour days without any stimulants other than the fear of what his children were doing while he was at work. Unfortunately, he soon had to give up decaffeinated coffee and sugarless candy with cyclamates since both caused laboratory mice to develop an incurable yen to listen to music from the Grand Funk Railroad.

Left with a diet of fruits and vegetables, he was lean and trim. Until he accidentally stumbled across a protest by an environmental group which complained that the use of pesticides on farm crops was a greater health hazard than the bugs the pesticides were supposed to kill. Even the city's polluted water couldn't clean off all the pesticides. That's also when he stopped taking showers, and merely poured a gallon of distilled water over his head every morning.

For weeks, he survived on buckets of vitamins. Then, after reading an article that artificial vitamins shaped like the Flintstones caused dinosaur rot, he also gave them up.

The last time I saw Bouldergrass alive, he was in a hospital room, claiming to see visions of monster genetic tomatoes squishing their way toward him while mumbling something about cholesterol and high density lipoproteins. Tubes were sticking out of every opening in his emaciated body, as well as a couple of openings that hadn't been there when he first checked in. Shortly before he died, he pulled me near him, asked that I write his obit, and in a throaty whisper begged, "Make sure you tell them I died healthy."

JUNE 1994

The Stupid Season

Less than a month before high school graduation, the seniors are celebrating impending unemployment by skipping class, as if they haven't skipped enough times already. Nevertheless, for the traditional "Skip Day" they're skipping off to party somewhere, while the smarter seniors are skipping both classes *and* the party.

It's the beginning of the annual Stupid Season which will end with at least a fifth of the nation's 13 million teenagers attending proms, that ostentatious cap to a 12-year flirtation with education. Boys once wore sports suits, and girls wore dresses to the prom. But, with the schools still hung over from the "Affluent '80s," it's now a rented tux and mirror-shine shoes for about $100, or a new $150-$500 prom gown which will soon hang in the closet as moth food. About 98 percent of the girls buy a new dress, averaging $200, according to *Your Prom* magazine.

Haircuts are about $10 for guys; perms are $25-$100 for the girls. Add in another $100 for nail polish and fake nails, lipstick, mascara, perfume, and new hose. For that special splash of color, you need a $5 carnation boutonniere for the guy, and a $20 orchid corsage for the girl—or, maybe a $60 bouquet of a dozen roses.

Some boys rent new cars; almost half, says *Your Prom*, will arrive in a $200-$250 a night limousine in vain attempts to impress whoever it is they believe they must impress. The rest apparently wash, wax, and vacuum their own cars, relatively recent pretend high performance red or black models which they park over four intersecting spaces so no one can hit their turtle-wax shine. To support the turtle, they work 20-30 hours a week at a minimum wage dead end job. When anyone asks why they don't just quit and spend the time studying, or become involved with extracurricular activities, they say they need the job to support their car and stereo.

A long time ago, the boy's extended family worked on a special meal for the prom couple. For some, circumstances allowed a nice dinner at an inexpensive restaurant. Now there's often only one parent in the house, and dinner is about $20-$25 each.

Juniors once decorated the gym for the prom. Now, it's held at the country club or the "Sweet Magnolia Room" of the high-rise hotel. Add tickets at $20-$35 per couple, and prom pictures for about $25.

There once were live bands. Now, DJs stuff CDs and tapes into a 125 watt supersystem. It makes no difference since most of the seniors haven't a clue of what music is anyhow. (After all, they didn't have time to join the band or chorus since they had to work that dead-end job to pay for that turtle-wax car.)

Sometime during the evening, in a country which says it doesn't believe in royalty, a king and queen are announced and, like the monarchy in England, no one seems to know what it is they're supposed to do.

Many schools have figured out ways to make the prom night extra special by extending it until morning. They take the students' car keys, bring in a lot of donated food, and hold hourly raffles of everything from CDs to a car. But, for most, the prom is over by 11. It means another $10 for after-prom dessert and drinks, or a $5 "contribution" at a back-woods keg party. For some, it means another fifty or so bucks for the after-prom motel room. (Somewhere in all this is the second mortgage most families take out so their sons and daughters don't feel emotionally deprived.)

Let's move on to the graduation itself. The Supreme Court has ruled that prayers at graduation violate the Constitutional provisions of church-state separation. However, many school administrators apparently don't believe the Constitution applies to them. Tossing out excuses that would embarrass even an obtuse 10th grade civics student, the educational leaders claim that the Supreme Court ruling "isn't really final," that it "violates our freedom," or that "we've always done it, and we're not going to change now." Of course, their arguments sound strangely like ones many school officials said after the Supreme Court ruled segregation unconstitutional. At the same time they're making plans to defy the Constitution and the Supreme Court, the administrators are throwing thousands of students into detention for defying even the most insignificant rule imposed by the schools themselves.

After prayers, the graduation speakers talk about how good the graduates' education has been, what glorious futures they have, and how they're the movers and shakers of tomorrow. Few point out the world they are about to move and shake has yet to reduce wars, poverty, crime, the health care mess, racism, sexism, and all the other -isms. No one says the SAT scores have been dropping for the past two decades, that 10 percent of the 2.5 million graduating seniors are functional illiterates, the economy is in a sinkhole, and that exploitation of the work force is a reality far more ominous than losing the basketball championship. Sadly, until the graduates get their wake-up calls, they will continue to polish their cars and overspend their allowances on self-indulgent frills.

MAY 1993

All the Beautiful People

From a pool of almost six billion contestants, the divinely-inspired *People* magazine editorial staff have once again chosen the 50 Most Beautiful People in the Whole Wide World, and duly etched them into the public mind by large color pictures and cutesy capsulized biographies. And, once again, it's also time for my annual *People* debunking. So, let's see if the people who put out *People* have learned anything in a year that saw wars, famine, and O.J. dominate the news.

Two years ago, the geniuses who make up *People*'s editorial staff decided that 23 of the world's 50 most beautiful people (46 percent) were actors or actresses. Last year, *People* decided actors and actresses were still 46 percent of all the beautiful people. This year, beauty explodes all over the screen as 18 actresses and 11 actors, 58 percent of all beautiful people, made the list.

Two years ago, *People*'s editors decided that seven singers were among the world's most beautiful list. But last year, there were only three, and this year only four, three of them men.

Models also lost position on the "elite 50." Two years ago, *People* decided that six models, all women and no men, were among the whole world's 50 most beautiful. Last year, there were still six women on the list, but now there were two men, apparently for gender balance. This year, it's one man and two women.

Thus, almost three-fourths of the beautiful people are in the entertainment industry, leading me to believe that appearance counts more than raw native talent in American life. Or, at least, we commoners demand people in the entertainment industry to be beautiful. Interestingly, almost three-fourths of the full-page ads in the section feature "beautiful women."

Two years ago, five athletes (one woman and four men) made the list. Last year, year, six athletes (two women and four men) made the list. This year, only three athletes, all of them men, made the list.

For the fourth year in a row, celebrity lawyer John F. Kennedy Jr. is a beautiful person. Perhaps the editorial staff could just give him a Loving Cup and let him retire as an Undefeated Beautiful Person.

The first year I compiled the list, I complained that no journalists or teachers were selected to be beautiful. Apparently, that message got through to *People*'s New York/Hollywood mentality. There still aren't any beautiful teachers, but there are now beautiful journalists. Last year, *People* selected former journalist and current Vice-President Al Gore. In addition to our Vice-President, the beautiful people last year included a husband-wife documentary film team, an NBC "Today" show news anchor, and an ABC "Wide World of Sports" interviewer. Of course, some cynics might suggest that the two network beauties are really entertainers not journalists. So this year, possibly hoping to avoid my probing pen, *People* actually found a real journalist, 61-year-old Gloria Steinem, who should be honored to be recognized for being beautiful, but embarassed by being included in a racist, sexist list. Nevertheless, for balance, People selected a male correspondent from NBC's "Today" show.

Teachers, alas, still didn't make the cut this year. But, they need not worry about it. Neither did Miss America, Miss USA, Miss World or, for that matter, Miss Crustacean, Ocean City, New Jersey's, tribute to beauty.

Personally, I would have selected fat and balding Charles Kuralt as one of the world's most beautiful people. His words and truth are far more beautiful than all the vacuous heads who made up the list that takes up 60 full pages of a national magazine.

The elderly were finally recognized last year as beautiful. Of course, *People*'s idea of elderly were 69-year-old Paul Newman, 57-year-old Barbara Babcock, and 53-year-old youngster Faye Dunaway. This year, the "elderly" included Steinem and 51-year-old Queen Silvia of Sweden.

For the first time, *People* tried to reflect reality last year by selecting 5-foot-11, 180-pound size 14 model Emme as a beautiful person, and pointing out she is a top-star in the "burgeoning large-size modeling industry." Here's a shock, *People* editors. Size 14 isn't fat! More than half of America's women are at least a size 14! If *People* wanted to highlight the reality that beauty comes in all sizes and shapes, it might have selected plump actress Shelly Winters. Nevertheless, apparently not wanting to set a trend by letting us commoners think beauty comes in all sizes, there are only modishly thin and ultra-thin selectes this year.

Now, let's look at racial and gender balance. About a half billion people in the world are Black; about 1.9 billion are Asian. Two years ago, *People* could find only nine beautiful people in the entire world who weren't Anglo-Saxon white, and eight of them were Americans, most of whom photograph as light-skinned. Last year—what a surprise!—only nine of the list, eight of them Americans, also weren't white. This year, in a leap into almost social relevancy, People found all of 13 who weren't White. Can't have too many of them Blacks and Hispanics cluttering up those beautiful lists. Interestingly, the ethnic distribution in each of the professions is also almost identical, including the reality that the previous two years, the only Asian represented each year was an actress, each of whom was born in China.

This year, in a major shift to recognize that there may be parts of the world other than Manhattan and Hollywood, *People* claims that a quarter of the most beautiful are "international." But, most of them are American residents, and all but two are from western Europe. Obviously, there aren't any beautiful people living in five of the world's seven continents.

At least there's gender equality. For each of the past three years, as if some quota was etched into *People*'s pancake-and-cream supply, there have been 26 women and 24 men, about the U.S. average.

To its credit, *People* editors, probably as an after thought, might have been concerned about why no common people made the list. So, two years ago, the editors went "cutesy," and found two cities in Kentucky—Lovely and Beauty—and awarded "booby prizes." Perhaps the editors prefer to think of it as "honorable mention," but "booby prize" seems more appropriate. Last year, *People* found no "commoners," so this year, they searched their loading docks and found nine UPS drivers—mostly men from New York City, Los Angeles, and Florida—to feature in the "booby prize section."

Nevertheless, Lovely, Beauty, and UPS aside, we can only conclude that *People* editors believe almost all the beautiful people of the world are upper-class White entertainers from America, almost all of whom have entourages of at least a publicist, business manager, and agent.

Like all people, we in journalism tend to report about, be attracted to, and understand people and ethnic groups that are like our own. And, for the most part, we are White middle-class, sometimes even upper-class, college graduates who talk a lot about equality, but look, act, and dress as if we are part of the establishment we report about. In fact, until the 1960s, most reporters were White males. If we see only certain groups of people all the time—and *People*'s editors apparently see only certain groups of people—then we will report only about those people, leaving everyone else as invisible as the billions of beautiful people all over the world who didn't make the list.

May 1995

There She Is, Miss-representing America

It's time for the annual Miss America Pageant, and once again America will probably be represented by a never-married 21-year-old college student who's 5-foot-5, 119 pounds, believes in God and family values, and claims she wants to save the whales and protect the homeless. The only thing we don't know about her is what shade of blonde she'll be.

The contestants all say they're in the pageant to earn a college scholarship and, since they're so beautiful, they want to become broadcast journalists. No one says she wants to be a waitress, small business owner, auto mechanic, or even to meet some hard-body with perfect razor-cut hair, get married, and have 2.3 genetically-perfect children.

In 1881, the *Police Gazette*, a cross between the *National Enquirer* and *People*, reported that following a nationwide search, Louise Montague, a burlesque actress and singer, was awarded $10,000 after being judged to be the most beautiful woman in America. She was described as having "a beautiful light complexion [with] a mass of wavy dark chestnut hair." The magazine also reported she was "of medium height [with] a full and symmetrical figure. Her weight is 147 pounds." At the turn of the century, America's sex symbol was singer/actress Lillian Russell. She weighed almost 200 pounds. Margaret Gorman, who was named the first "official" Miss America in 1921, had a rather flat 30-25-32 figure. By today's Miss America beauty standards, none of them would have made it into the finals of the local Miss Rutabaga contest.

Although swimsuit and evening wear competition each count for only 15 percent of the total, it seems the contestants spend 90 percent of their time perfecting, sometimes with collagens, the pouty-lip look that is "in," and spend hours applying makeup to every available millimeter of emaciated skin. There's vaseline to glisten, tape to firm up, and rhinestones for reflection. This year, for the first time, the contestants will have to do their own hair and makeup. As important as these skills are, it might be more impressive if they went to the flooded Midwest and stuffed sandbags along the Mississippi.

There's also a new wrinkle in the swimsuits. Instead of being honest and saying that American men want to see women in swimsuits, the organizers now claim in a puffed-up news release that the swimsuit competition is to allow judges to determine "each contestant's perseverance and self-discipline in maintaining a physically fit and healthy body." Maybe next year, the contestants might run laps, or be required to submit results of blood tests and X-Rays. Perhaps, psychologists and physicians will replace the celebrity judges. Nevertheless, just so we can see those physically fit bodies, the contestants wear one-piece suits that firm and raise—but not too much; after all, we don't want to offend Aunt Myrtle and the Citizen Decency League. Those swimsuits are as much designed for splashing in the ocean as mackerels are meant to wear lipstick and eye liner.

Because they've become sensitive about image, the organizers tell us, trying to keep from smirking, that the mind and talent are the most important parts of the pageant. And, to prove it, the organizers have even programmed their telephones to answer, "Miss America Scholarship Program," apparently to make us believe that beauty has brains and will shortly complete that doctorate in biochemical engineering.

At least half the contestants in the 70,000 or so preliminary beauty contests sing "The Impossible Dream." What's impossible is they actually believe they can sing. If they don't sing, they play the piano, violin, or flail their arms and legs and pretend it's "modern dance." A few even pretend they're actresses and read a monologue from a play they never knew existed until someone from the pageant told them they had to develop a talent real fast or else lose the right to publicly proclaim they believe in the environment. What we don't see are writers, artists, or sculptors; we see no one who can make quilts, design dresses, or is a fine comedienne. If the organizers truly believe that talent is 40 percent of the score, then why weren't Bette Midler, cartoonist Cathy Guisewite, Grandma Moses or Roseanne ever in the finals?

Now, just in case they have little talent, contestants can redeem themselves in interviews. The organizers proclaim that the "newly expanded private twelve-minute session," worth 30 percent of the total, is to evaluate contestants on "such subjects as world affairs, state and local politics, personal interest" and, of course, "interpersonal relationships." During interviews, the contestants recite carefully prepared speeches, sweetened by voice coaches who have studied interpersonal communications, and can now cover any possible question with the same prepared answer.

Because this is a *pageant*, not a beauty contest, Miss America contestants must present their "platforms," their goals to change the world should they be elected beauty queen for a year. In fact, the Miss America mandate, say the organizers, is to encourage young women to explore the relevant social issues of their times and to excel in arts, science, communications or any area of inquiry that inspires their interest and devotion." But, how many contestants are going to be allowed to have pro-choice or pro-life platforms? How many will forcefully argue that corporations are exploiting foreign labor? Know anyone past preliminary contests who contends that lesbians should be allowed to serve in military combat units?

If we truly want someone who embodies America, then why not bestow the crown upon attorney general Janet Reno. She's single, has immeasurable talent in a field that pageant organizers don't include in their inconsequential world, can certainly out think any of the contestants in interviews, and is far more beautiful a person than most the finalists. It took the organizers more than a half-century to figure out that there were Blacks, Hispanics, and Asians in America. Until the organizers recognize that beauty comes in many packages, then no matter what PR and glitz the organizers use to sugarcoat their contest, the reality is that it's still just another beauty contest that excludes most women from consideration.

AUGUST 1993

In an inspired moment of public relations, Pageant officials in Summer 1995 declared they would allow the American people to vote on whether or not to keep the bathing suit competition. It was a no-lose plan. No matter how the public voted, the timorous officials could merely claim they were following the will of the people, while reaping the PR deluge that fell with the creation of the con-

test. On the night of the Miss America Pageant, almost one million viewers called in to one of 11,000 phone lines; by a 4-1 margin, they said they wanted to keep bathing suits in the competition.

Jeanetics and the Cool 'Wannabe'

It's the end of Summer, and time to fork over what's left of my salary to buy a few pairs of new blue jeans for my sons. Not just *any* pair, but ones that are faded and torn in the manufacturing process. It's some kind of a "fashion statement." More than 28.5 million teenagers a year spend about $18 billion a year, half of it in August, for clothes. Parents and relatives spend another bazillion or so dollars.

I usually wait until the last week, hoping there will have been a massive switch in advertising strategies, and jeans won't be the one item that identifies everyone who wants to appear to be young, cool, and "with it."

Just about everyone is wearing jeans. Students and teachers. Patients and doctors. Lawyers, cops, and criminals. In upscale Manhattan, the Yuppie crowd wear jeans with a blue blazer. In middle America, reporters wear jeans with inexpensive shirts and ties. Wannabe rappers wear baggy jeans, and wannabe sex kittens wear low-cut form-fitting jeans with strategically placed designer holes cut in places that only the size 3 models the companies hire as shills would be happy with their placement.

Cowboy president Ronald Reagan wore jeans on his Santa Barbara ranch while taking one of about 10,000 vacations from his demanding eight-hour a day work schedule in Washington. Nuclear engineer and former president Jimmy Carter wears jeans since he was a peanut farmer and later became active in the Habitat for Humanity program of building houses for low-income families. During their cross-country campaign of 1992, Bill Clinton and Al Gore, those affable Ivy League lawyers, wore jeans.

On labels wide as barn doors, to be displayed on bodies nearly as wide, are free advertising for Calvin Klein, Gloria Vanderbilt, Liz Claiborne, Ralph Lauren, Sergio Valente, Vidal Sasson, and Guess? (no, I don't know Miss Guess's first name!) Even Georgio Armani, manufacturer of astronomically-priced upscale suits, ties, and shirts, is manufacturing astronomically-priced jeans.

About 450 million pairs of jeans, made from about 9,000 tons of denim, were sold world-wide last year, accounting for about $9 billion in sales. The leader is Levi Strauss Co. which in more than a century has sold more than 2.5 billion pairs of jeans, and currently has a 20 percent share of the market, ahead of Lee and Wrangler, both VF Corp. divisions. The annual advertising budget solely for rivets-on-the-pockets, buttons-on-the-fly Levi's 501s, the tight-fitting straight-legged, narrow-ankle, rump-enhancing best-selling jeans in the world, is about $20 million a year.

When Levi Strauss made his first pair of jeans in 1853 during the California Gold Rush, he didn't worry if *GQ* or *Vogue* gave it their seals of approval. The pants, first known as waist overalls, were strong and durable. Miners liked them. Farmers liked them. The poor, who couldn't afford expensive wool slacks, liked them.

In the early 1860s, Strauss changed the original brown canvas overalls to an indigo-dyed denim fabric. Later, he added suspender buttons instead of belt loops. In 1872, almost two decades after Levi's were first made, Jacob Davis, a tailor, added rivets to pockets so they'd be stronger. Other innovations included stitched patterns on the backpockets the following year, and a fourth pocket (the watch or coin pocket) in 1890. A fifth pocket (a second one in the back) was added 15 years later; belt loops were added in 1922, and the red tab in 1930. Even with all these inventions, for more than a century the rest of America never wore jeans, lest others believe they were miners, farmers, or poor.

Then, in 1957, Elvis wore a pair of black denims in "Jailhouse Rock," and the rock and roll generation began demanding jeans. A decade later, the hippies started wearing bell-bottom jeans, and a decade after that, Calvin Klein merged sex and denim.

It was in 1980 that 15-year-old starlet Brooke Shields seductively asked, "You know what comes between me and my Calvins?" She then cooed "Nothing," and jeans became the aphrodisiac all of America wanted.

Soon, it seemed as if every clothing manufacturer was into rounding up the denim supply and painting it onto their models. Women who could pour themselves into a pair of designer jeans with nothing hanging over were soon admired in our society; they apparently mastered the art of self-restraint to be thin enough to be uncomfortable. Size 14 women were trying to merge into size 8s, believing they'd look sexy being strangled in denim.

Sex continued to pour out of America's clothes in the 1980s as stonewashed and faded ripped and torn jeans, which exposed flesh while deceiving America into believing the appearance of an active life complete with peek-a-boo sex was better than the real thing, became America's fashion statement.

However, just to make sure they don't miss any of the market, several companies began making roomier jeans, with VF's Lee division coming up with a series of ads for Easy Riders, targeted to the 25-44 year old market, that showed women struggling to get into the tight-fitting jeans. "If you really want to relax," suggested a Lee print ad, "maybe it's your jeans that should loosen up." Another ad asked, "Need a little more room in your jeans? Try Easy Riders from Lee. The brand that fits."

Wrangler targeted readers of *Cosmopolitan*, *Glamour*, and *People* with its campaign—"Our philosophy on fit: Jeans should take his breath away, not yours." Wrangler, a rugged wear line targeted to the country music market, is sold to the theme of "Most country stars wear Wranglers." For one of its major campaigns, Wrangler hired country singer George Strait to model the Wrangler brand, and told the public, "A western original wears a western original." The company advertises heavily on The Nashville Network, and spends almost $1 million of its $10 million advertising budget to sponsor the Professional Rodeo Cowboys Association, placing its print ads on almost every available space, including barrels in barrel races. Other advertising is placed in *Horse Illustrated*, *Western Horseman*, and *Quarter Horse Journal*, as well as *Field & Stream*, *Sports Illustrated*, *Rolling Stone*, and *People*.

Although continuing to emphasize the poured-into-denim look for both men and women, Levi's countered with the looser fitting 560s. But, to hold the market on the "original jeans," Levi's increased its promotion for the 501s on FOX, MTV, and Nickelodeon, with the "got to be real" campaign to make sure the consumer knew that the jeans with the "button fly" was the original, not some baggy imitation.

During the 1990s, Calvin Klein, leading the fashion industry with sexual-oriented advertising, continuing to sell sex and jeans, posing pliable and emaciated nude women draped over brawny men wearing nothing but—what else?—their Calvin Kleins. In 1991, Calvin Klein spent $1 million just on advertising space to run an erotic 116-page photo spread of naked men and women, jeans, Harleys, and tattoos in 250,000 copies of *Vanity Fair*'s normal 850,000 circulation run. The section, Klein told *TIME* magazine, "is a fantasy about a rock concert. You see the band onstage, backstage, after the show. The wild and crazy groupies. The people living in the motorcycle world. It's about excitement. Hot and sweaty rock 'n' rollers who wear nothing but jeans and skin. It's about denim. People love it." That year, Calvin Klein spent $10 million on advertising, most of it mixing jeans and sex.

The following year, Calvin Klein dropped jeans entirely in one of its ads, posing rapper Marky Mark on posters wearing nothing but an alluring smile and a pair of undershorts. In the Summer of 1995, Calvin Klein shocked America once again by running an ad campaign featuring panels of sexually-provocative pubescent boys and girls. He never revealed their ages, but dropped the campaign a couple of months later against strong public outrage.

The next fashion statement probably won't be denim underwear, although consumers can complement their jeans with denim vests and jackets, or make an "alternative lifestyle" statement with denim skirts, jumpers, or coveralls.

And that's why I'm again saying goodbye to my paycheck, resigned that for another season jeans are, apparently, what everyone else wants whose mind has been fashioned by multimillion dollar advertising campaigns that survey the market then tell us not only what we want, but what we need.

AUGUST 1995

Sex and the Single Beer Can

I

The First Commandment of successful advertising is as simple as some of the campaigns—Sex Sells.

Car manufacturers paint two-door convertibles fire-hot red. Jeans manufacturers fleece the public to believe there's nothing sexier than someone in tight jeans. And beer companies have spawned bikini-clad lithe spirits to con cigar-chomping overweight never-hunks in grease-stained sloppy T-shirts to swill one or two six-packs every Sunday afternoon in front of a football-filled television set and believe they're the sex-machines all America is looking for.

Throughout the country, we're exposed to Blondes in Bikinis. For variety, there's Blondes in Tank Tops, Blondes in Halters, Blondes in Unbuttoned Short Shorts, Blondes in Cowboy Hats, and Blondes in Paint-Them-on-My-Body Jeans. When the beer companies decided they should show a little more "multi-culturalism" in their advertising, they added brunettes and redheads—and put them into bikinis, tank tops, halters, unbuttoned short shorts, cowboy hats, and jeans. Three decades after the beginning of the civil rights era, the breweries finally found white-looking size 5 Black women to promote beer—and gave them the opportunity to stretch out on white sand and lure customers into an idyllic setting of sun, surf, sex, and beer.

During the holidays, the models wear clinging dresses with

hems somewhere about waist level—red and white for Valentine's Day; red, white, and blue for Independence Day; orange for Halloween; furry red and white for Christmas; and green for St. Patrick's Day. Michael Shea's places a blonde in a slinky green dress cut above mid-thigh, a shamrock, and a beer bottle to entice the public to try "a naturally pure golden lager." Miller Lite, the nation's third best-selling brand behind Budweiser and Bud Lite, places three women in short green dresses and asks America's sex machines to "triple your Irish."

On billboards, posters, and banners, the message is clear. Drink beer and achieve great athletic and sexual satisfaction. Lured by the advertising, Americans are going to parties and bars to deliberately get drunk, apparently believing they can do things drunk they can't do sober—like vomit all over their dates.

Upon such dreams, Anheuser-Busch, Miller, and Coors, all of which connect sex and the consumer, accounted for 80.3 percent of the $51.1 billion spent for beer last year, and about 89 percent of the $705.9 million spent on advertising, about 85 percent of it on television advertising, according to *Beverage World*, which tracks the industry.

Although the major beer companies effectively use humor, athletes, celebrities, and racecar drivers, somehow tying together "responsible driving" and drinking, their print advertising is more sex-driven. With sales of about $4.2 billion a year for production of about 43 million barrels, and an annual advertising budget of about $220 million, and 22.8 percent of the domestic beer market, Miller is the second largest brewery, behind Anheuser-Busch ($9 billion, 88.2 million barrels a year and 46.8 percent of the market) and ahead of Coors ($1.7 billion, 20.2 million barrels and 10.7 percent). One of Miller's more popular campaigns was to promote its genuine draft product by showing a dress and shirt on a clothesline. In case the consumers missed the subtlety, it also has posters of models in skin-tight swim suits, one of whom is caressing a Miller Genuine Draft beneath a waterfall.

Waterfalls or showers are the setting for almost every brand in the Anheuser-Busch, Miller, and Coors lines. Coors, with an annual advertising budget of about $120 million, extols cool, clear water—and string bikinis, cleavage, and beer. Its slender D-cup women are often seductively stretched out on beaches, purring about "real beer satisfaction."

In case a few rugged he-men aren't into beaches, there's macho-man motorcycles. One Coors poster shows a lady in short shorts, a halter top, and a leather jacket standing next to a Harley-Davidson. For the sophisticated beer-guzzler, there's a poster of a model in black heels and short black cocktail dress about to straddle a Harley Softail. Her fuel of choice? Coors Extra Gold. Possibly unwilling to pay a six-digit licensing fee to Harley-Davidson—or having already spent its budget on $10,000 a day models—Michelob Dry, an Anheuser-Busch product, has a model in black heels and black cocktail dress merely caressing a beer bottle. For "variety," Budweiser, also an Anheuser-Busch product, has a brunette in a short red dress sitting seductively on steps. Nevertheless, most of the nation's major breweries have copied each others' "sophisticated" look to promote their "premium" beers, hawking their not-so-subliminal attempts to make people believe that beer and "class" are synonymous.

Not so sophisticated is Colt .45 from G. Heilman, the nation's fifth largest brewery (8.3 million barrels, 4.4 percent of market share), which promotes its product by having a seductive woman in a short silky blue dress down on her hands and knees, her cleavage and come-hither look telling men, "Everytime." Keystone State, a Coors product, uses blue-and-white striped one-piece swimsuits on its models to promote the brand's label color. Killian's, another Coors brand, uses red. Its model is a tall redhead in a white bikini who exalts salivating men to "try a tall cool red one." Evoking an Irish theme, Mickey's features a large-busted model in a body-hugging green swim suit; between her spread legs is an obviously-phallic Mickey's wide-mouth bottle.

Missing from beer advertising is "beefcake," male models in muscles and Speedos. Although the average per capita consumption of beer is 56 gallons a year for men who drink beer, according to *Impact Magazine*, women are now about 35 percent of all beer drinkers, almost half of all Lite drinkers, and consume an average of about 14 gallons a year. When the breweries and their ad agencies figure out that women respond to beefcake, they'll supply it. But, for now, women apparently are lured to beer for other reasons.

II

The corporate spokesmen for the nation's largest breweries all talk about "responsible advertising," and how they deliberately avoid targeting the younger generation. They say their advertising isn't sexist. However, breweries have been using women as lures in beer advertising since the late 1800s. The first ads featured pictures of the breweries and their logos, or cartoony bartenders or families enjoying beer. However, by the early twentieth century, the companies began using women in their advertising. Most ads featured merely the face, heavily made up. The more adventurous featured women from the waist up.

Most of the women had long flowing hair, possibly a subliminal message that the models had "let their hair down" for the male beer drinkers. Images were lithographed onto beer trays, coasters, and posters, and etched into newspapers and magazine advertising. Often, the breweries combined women, beer, and the latest autos, especially during the first two decades of mass production. For several decades before prohibition, Budweiser had featured the angelic-appearing "Budweiser Girls" on trays and related advertising media. Marathan featured a "bobby-soxer" in a tight sweater and short skirt suggestively posed next to a bowling alley, with the slogan, "Strikes You Right." Rheingold brought national attention to its product with the annual Miss Rheingold contest from 1940 to 1965. Rieger & Gretz Brewing Co. of Philadelphia, like many other breweries, combined the "angelic" with the sexual, featuring women in translucent gowns for annual calendars.

With the "sex revolution" of the 1960s, and increased competition, the major breweries lost whatever was left of their ages of innocence, and combined suggestive slogans and nearly-naked women with the "come hither" look.

Stroh's, the nation's fourth largest brewery with sales of about $700 million a year for 12.1 million barrels, and 6.4 percent of the domestic beer market, tweaked the public in 1991 with a campaign for its Old Milwaukee Brand that featured the Swedish Bikini Team, four platinum-blonde women who parachuted into a camp of men. The campaign was supposed to be a parody of the sex-based campaigns of the other breweries, but was killed after one season when several Stroh's employees protested the use of sexism to parody sexism.

Disregarding Stroh's experience, Coors unleashed the Artic Angels in 1993, three blonde women in thigh-high black boots and body-tight black body suits to promote Artic Ice. However, Janet Rowe, Coors manager of corporate communications, claims her company hasn't used sexist advertising "for some time." Not even the Artic Angels? That's a "local promotion discretion," says Rowe. Almost in the same breath, she admits the company "has not developed any formal new policy on women in bathing suits." Nevertheless, although Coors' TV advertising is using more humor and less sexism, the bathing beauty posters are still prevalent in beer stores.

Even against such formidable competition, Stroh's stuck by its earlier decision to reduce sexism in its ads, although one poster shows a model wearing a Chicago Bears jersey and biker shorts. "Our whole campaign now," says Burke Cueny, Stroh's marketing director, is "more taste, more character. We're interested in featuring our beer in everyday situations as opposed to a fantasy-setting agenda."

With gross sales of about $9 billion a year, and 46.8 percent of the domestic beer market, Anheuser-Busch is well ahead of the field. Its Budweiser brand, with 41.4 million barrels brewed a year and 22 percent of the domestic beer market; and Bud Lite, with 16.5 million barrels and 8.7 percent of the market, are the two best-selling brands in the country. About 90 percent of its $280 million a year advertising budget goes into television. However, unlike other major companies, Anheuser-Busch doesn't advertise on MTV because, says the company, "MTV skews too young." Although Anheuser-Busch still fuses women, sexual suggestion, and beer, the print ads tend to be less "racy" than those of the other companies, although they still feature women in short and tight dresses or modest bathing suits. The company began cutting back on the bikini crowd in the late 1980s. "There are other ways to convey the message," says one of the conglomerate's spokesmen.

Genessee, distributed primarily in New York and Pennsylvania, doesn't connect sex with beer. With production of 1.9 million barrels a year, and annual gross sales of $200 million and a $2 million ad budget, Genessee is the seventh largest brewery in the country. Hugh Crossin, Genessee's director of advertising, says even with a healthy advertising budget, the company has no plans to succumb to lust. With only a one percent market share, Genessee, says Crossin, "can't compete with the giants, so we must establish our own identity."

Yuengling, the nation's 13th largest brewery, with gross sales of about $200 million a year, sells about 90 percent of its 275,000 barrel annual production in eastern Pennsylvania. Like Genessee, Yuengling also avoids sexist advertising. "We haven't had to resort to those tactics," says Dave Casinelli, vice-president of sales and marketing for the nation's oldest brewery. He says Yuengling sells "on the basis of consistent qualities. We just can't compete with the majors."

Competing with the Top 3 companies would be economic disaster for most of the other breweries. The cost to produce one poster for distribution to beer stores and customers could be $100,000. With few exceptions, almost all of the $7.4 million advertising budget for breweries not in the Top 10 is directed less on image, more on product and substance.

"We don't need sex to sell our beer," says Dan Straub, president of Straub Brewing, St. Marys, Pennsylvania, which promotes its all-natural beer on the basis of taste, quality, and being "honestly fresh." Word-of-mouth, posters, and "a lot of community service" make Straub known in western Pennsylvania.

However, if Straub would stop worrying about quality, slash a few hundred thousand from research and development funds, cut back on the quality of his ingredients, then triple his $100,000 a year ad campaigns, he could hire a couple of models with IQs the size of their busts and put them into G-strings to coo to the public, possibly at a community service wet T-shirt contest, "I like to be Straubbed." Maybe then the nation's 46th largest brewery could sell more beer.

JUNE 1995

Facing a McBlimp Attack

Beneath a clear blue winter afternoon, I was lying face down on the parkway outside city hall. On top of me, cursing and screaming they'll never take us alive, was Marshbaum. The last thing I had remembered before being hit with a flying tackle was looking up. So, I looked up again.

"Stay down!" Marshbaum barked.

"Snipers?" I fearfully asked. When you're a political satirist, you never know who you may have offended.

"Blimp," whispered Marshbaum ominously.

"The Goodyear blimp is after me? It's as American as junk bonds."

"That's what they want you to believe. Didn't you ever wonder what they *really* do over all those football games? They're gathering intelligence."

"What can they learn at a football game?"

"Enough to know that 50,000 drunks wearing hog noses to cheer for the Redskins aren't going to be able to provide much resistance to an all-out invasion."

"Goodyear is going to invade Robert F. Kennedy stadium?"

"Goodyear is on our side," he said. The problem is the other blimps. The Fugi blimp is probably in luke-warm pursuit right now."

"We're going to war with a company that makes film?"

"A *Japanese* company," said Marshbaum smugly. "When we defeated the Germans and had the only flying blimps, we had air superiority. The Japanese take our technology, improve it, and build their own blimp. Probably have a dozen quality work circles right now just on how to better vulcanize rubber. It's a worldwide conspiracy, Even the Scots are involved."

"Yeah," I said sarcastically, "the last thing an enemy wants to see is a tartan-design blimp floating into battle, staffed by Highlanders leaning out the windows, with their bagpipes blaring 'Comin' Through the Rye.' "

"McDonald's," said a smirking Marshbaum. "The Scots achieved parity with Japan when they convinced us that two-ounce McGreaseburgers were healthier than tofu and seaweed. Blimps are the future of warfare."

I had known that during World War I, blimps had been used as scouts. French and German blimps passed each other, their crews waving politely at the enemy. But then someone, probably loaded on dark beer or burgundy, threw an incendiary device at the other blimp, and soon blimps were in the war. Except for some coastal surveillance for subs during World War II, blimps weren't the attack craft military leaders once envisioned, mostly because even a blind squirrel with a nut could hit something that large and that slow. I demanded Marshbaum produce evidence.

Marshbaum looked around, saw that no one was watching, then took a crumpled piece of paper from his back pocket. "See!" he said, thrusting an unfolded news article at me. According to the paper, our government has been planning for blimp warfare since 1929. Beneath 20 square miles in Texas, the government has stockpiled more than 32 billion cubic feet of helium.

In 1960, three years after Sputnik was launched, the lighter-than-air Congress not only renewed the National Helium Reserves but increased the stockpile. Apparently, they were worried that the Soviets would achieve blimp parity and launch waves of 70 mile-per-hour inflated bags to sneak past our air defense perimeter.

By 1973, four years after Neil Armstrong walked on the Moon, Congress, which had been sniffing the helium reserves, decided there was finally enough gas underground. But since they had all this gas, they had to sell it. It didn't take many brain cells for Congress to enact a law to require NASA, and the departments of Defense and Energy to buy helium from the reserve. It costs $22 million a year to maintain the stockpile, and there's always a slight profit at the end of the year—except for the $130 million interest the government pays each year on the $1 billion it cost to buy all that air in the first place.

"Even if all this is true," I told Marshbaum, "there's no evidence that an imminent invasion is planned." But, at that point, chugging along at sub-warp speed, floated the sinister Shamu killer whale blimp, apparently on a search-and-destroy mission to rid us of the Goodyear blimp, and take over football fields and amusement parks throughout the country.

"It's only the beginning," said Marshbaum prophetically. "Only the beginning."
MARCH 1993

Downsizing Baseball's Problems

It's the end of the year, and there's some old business that must be disposed of before we can go frightened into yet another year. There are still atrocities in Bosnia, no cure for AIDS or most cancers, and children are still buying guns to kill other children. But first, we must solve the crisis in major league baseball.

The strike has already led to the cancellation of the World Series, and almost caused yet another international war since the Liechtenstein Leopards were on a winning streak and would definitely have made the playoffs this year.

The owners want a salary cap. They say it's because the player payroll has become excessive the past few years, with some players now being paid as much as $7 million a year. The owners claim that establishing a maximum payroll would level the playing field so all teams could be equal, thus giving the Colorado Rockies the opportunity to win a World Series and sell overpriced T-shirts and mugs with pictures of blizzards on them.

The players, however, claim the salary cap is a sham since the wealthier teams have won no more games than the less affluent, and the owners are already making excessive profits. In rebuttal, the owners say that only a few of them are making excessive profits, while the rest are only making outrageous profits.

The players rightfully point out that baseball operates as the nation's only legal monopoly, and that because of the failure of government to impose anti-trust sanctions, as it would against any other benevolent mega-maniacal-conglomerate, the government encourages the owners to treat players as chattel, to be bought and sold at will. The owners complain that free agency is a Communist conspiracy, and that the players are nothing but million-dollar cry-babies who are better protected under slavery than freedom. They say if the major leaguers don't like the playing for one owner, not only are there 27 others who won't let them work either, but that thousands of minor leaguers are just salivating for the chance to chew and spit in major league parks.

Even the federal government is tired of all the feuding, and has filed two unfair labor practices charges against the owners. The owners claim it was just "a misunderstanding" that led them and their stadium of lawyers to disregard what few laws did apply to them.

Because owners, players, and even federal labor mediators have been unable to solve the problems, it's finally time to turn to the source of all great wisdom—newspaper columnists. Fortunately, I have a solution.

Using private industry as a model, I propose there be layoffs, known among corporations as "downsizing." The owners could eliminate one player and his couple of million buck salary. That would allow a larger pool of money for the players, while also allowing owners to keep their profits. The downsizing would allow worker exploitation to continue without fear of the industry closing down and the end of baseball caps worn backward as a national fashion statement.

Analyzing the playing field, it becomes apparent that every player except one has a place to be. There's bags for 1st, 2nd, and 3rd, a home plate for the catcher, mound for the pitcher, and billboards on the fence to designate the three outfield positions. The only player who doesn't have a designated spot is the shortstop. The fans probably won't object to one less player since they didn't complain when airlines laid off mechanics, and hospitals laid off nurses. And they certainly don't complain when publishers and station managers order "downsizing" of the reporting staff in order to increase corporate profits.

To compensate for the smaller workforce, we could move the third baseman a few yards closer to second base, bring in the left fielder a few yards to help cover the hole at third, and require the centerfielder and rightfielder to cover more territory. Although the layoff would increase the workload, it would maintain owners' earnings-to-profits ratio, and continue to provide them with the income sufficient to be able to afford German-made luxury cars, overpriced country club memberships, and rent on apartments for their mistresses.

Equally important, it would preserve the inherent right of owners to continue to make idiotic statements to a fawning press corps, and would assure the continuation of the responsibility of all Americans to take out a second mortgage to afford stadium parking, admission, and the right to eat three buck overdone hotdogs and drink a four buck cup of warm beer.

DECEMBER 1994

Replacing Baseball

With several dozen Americans anxiously awaiting the Replacement Baseball Season, I thought I'd stop by Marshbaum Industries to see the latest sports equipment.

"Best replacement baseballs in the business," said an enthusiastic Marshbaum, throwing one my way.

"The stitching is coming off," I said.

"Gives the ball character," said Marshbaum. "Here's another one."

"Marshbaum," I said concerned, "these baseballs are different sizes."

"Well of course they're different sizes," said Marshbaum. "We make some a little larger than normal for the replacement batters, and some a little smaller than normal for the replacement pitchers. It levels the playing field."

"Why wouldn't pitchers just keep throwing the smaller balls?" I asked.

Marshbaum looked at me as if I had spent too much of my career inhaling press box odors. "Because in Replacement Baseball, the umpire will randomly select the ball. That way, it's fair."

"I see that you're also making gear for the umpire."

"Best quality cardboard that money can buy."

"Cardboard? Aren't you worried the umpires might get injured?"

"They're replacements!" said Marshbaum. "*Everyone* is a replacement. You lose a couple, there's more out there. Besides, with replacement pitchers, I'd worry more about being hit by a baseball if I were in the stands than behind home plate."

"But you're making shoddy merchandize," I protested.

"Of course it's shoddy!" said Marshbaum, annoyed that I didn't even grasp elementary baseball basics. "This is replacement merchandise. Besides, the owners aren't paying us as much, so we have to cut a few corners here and there. Want to see our felt catcher's mitts?"

I declined, but asked how the changes and reduced income affected his employees.

"After we laid off the ones who were here the longest and had the highest wages, we found several on the streets who were willing to work at any cost."

"You hired replacements?!" I said astonished.

"We prefer to think of them as individual productivity stations. It's a much better debt-to-profit ratio for us ever since we increased the working hours, lowered the wages, cut back on employee benefits, and told them they couldn't work anywhere else but here."

"Marshbaum, what you've done is to violate at least a half-dozen labor laws."

"This is baseball, ink breath. Laws don't apply to us."

"Are all the ball clubs buying your junk?"

"All but the Baltimore Orioles. The owner says he won't use replacement players and won't field a team against other teams that do. Some absurd reason like he's got integrity and believes in labor laws and the rights of the working class. As if *that* counts for anything in baseball."

"He has that right."

"Not if he wants to be an owner he doesn't. Right now, the other owners are planning to disenfranchise him for not acting in the best interests of baseball."

"Is there *anything* your company *isn't* making replacements for?" I asked.

"One thing," said Marshbaum. "What we really want to do is to make replacement owners."

MARCH 1995

Traded for Two Rookies, an Editorial Clerk, and a Future Draft Choice

Finstermeister was furious. He stomped up to the desk of the city editor and shoved the afternoon edition of the newspaper in his face.

"How dare you!" Finstermeister demanded, his rage filling the newsroom. The city editor peered over his stack of press releases, and with embarrassed compassion in his voice said, "Sorry."

"Sorry? Is that *all* you have to say!?"

"You know how reporters are," said the city editor. "They're a zealous bunch, and sometimes there's no stopping them when they sniff out a story. I was hoping to tell you before it hit the street edition, but . . . well, with all these press releases, I just didn't have time."

"You had time to trade me!" thundered Finstermeister, upset that after 15 years on the *Daily Tribune*, he was being traded to the *Morning Bugle* for two rookie reporters, an editorial clerk, and a future draft choice.

"It was in the best interests of journalism," said the city editor.

"The contract requires that with that much seniority, I must be allowed to approve the trade," Finstermeister said smugly. "And, I don't like Maine!"

"Don't you remember," the city editor reminded him, "you waived that right three years ago in order to get only a six-day week? It was just after your third child was born and your wife had two cops drag you home so you could also meet your first two children."

"But, Boss, I'm a darned good reporter."

The city editor shook his head sadly. "We don't need an environmental reporter any more. The readers want gossip about the stars. Fashion stories. More about investments and jobs and where to find good meals. Important things like that. We'll just add an occasional assignment to Smogbound's beat."

"I can do more than just cover the environment," Finstermeister begged. "My rookie year I led the league in rewrites, obits written under deadline pressure, fire alarms, and garden club parties. Five years later, I was a Pulitzer finalist for that series on hazardous waste dumping!"

"I admit you had some good years with us," said the city editor, "but you're not as young and as aggressive as you once were."

"Two years ago," shouted Finstermeister, "I took the Pulitzer for the series on design flaws and construction blunders on the Bellevane Nuke." Finstermeister was now inches from the city editor's face. "Oil! I did that series that proved Magnum Oil was price-fixing. I wrote about acid rain, the utilities' rate-gouging schemes, sewer run-offs, cloud seeding—"

The city editor cut him off. "Don't you understand? No one cares about the environment any more."

"I'll learn to make charts and graphics!" cried out Finstermeister, desperately. He paused a moment, "I'll go back to editing the home improvement tabloids."

"You're a star, Fin, and we need to get rid of you while you still have some value. Those sixteen-hour days are breaking you down."

"Was it my salary? I'll take a cut. I don't need $11,000 a year. That's too much for a reporter anyhow."

"That's generous, Fin, but for your salary, we can get two cub reporters and have pocket change left over. Besides, you fit the *Bugle*'s needs, and the people they're sending us fit ours."

"I was president of the Society of Professional Journalists!" Finstermeister blurted out. "Didn't that bring a lot of prestige to the paper?"

"It's the system, Fin. You knew it when you came here."

Finstermeister looked around the newsroom, searching for understanding faces, but all the other eyes were buried in newspapers, hiding from reality. There would be no help from his colleagues today.

"What about the readers?" asked Finstermeister. "A lot of people bought the paper because of my reporting."

"A month from now," said the city editor matter-of-factly, "they'll forget your name. They'll find other heroes." Suddenly, the city editor became menacing. "And don't try retiring early, or jumping the team once you're at the *Bugle*. The commissioner frowns on things like that."

Sadly, Finstermeister shuffled his feet, resigned to his fate. "When do I report?"

"Take a vacation first and report in three days. The *Bugle* likes its reporters fresh." He paused a moment. "One other thing, turn in your style manual before you leave."

JUNE 1994

Laying Off Marshbaum

It was late Friday when I somberly called Marshbaum into my office. There was no way to break it to him gently, so I laid it on the line.

"Marshbaum," I said, "I'm going to have to lay you off."

"What'd I do wrong, Boss?" he asked, wounded.

"It wasn't anything you did," I said reassuringly. "It's what you are."

"I'm your foil," he said. "When you need someone to come up with dumb ideas that have an edge of truth, I'm it."

"That's true, but you're also a white male, and I've used you too much."

"That's my job!" said Marshbaum. "I'm the one who makes you look good."

"You're the one who's going to cause me problems with diversity committees," I replied.

"You telling me that because I'm a white male you have to lay me off?"

"I have no choice," I said showing him a study commissioned by the Screen Actors Guild and the American Federation of Television and Radio Artists, and conducted by a research team at the University of Pennsylvania.

According to the research, of more than 19,000 speaking roles in almost 1,400 shows over 10 seasons, women comprised less than one-third of all speaking parts. The study also said Latinos, Native Americans, the elderly, the disabled are almost invisible on TV, and lower class persons were represented in only 1.3 percent of major characters in prime time although there are ten times that many in the general population. Even more incriminating, the study reported that almost two-thirds of all on-air TV journalists are men, and that four-fifths of both those cited as authorities and those whom the networks consider newsmakers, are men.

"But that's TV," said Marshbaum haughtily. "We're *print*."

"Makes no difference," I replied. "If they can target TV, they'll find us next."

"Even if the networks believe the report, they won't do anything about it."

"And what makes you think that?"

"Because TV's never done anything socially responsible," said Marshbaum smugly.

"I still have to lay you off," I again told him.

"Sure," Marshbaum said cynically, "then you'll up and move the column to Mexico and take advantage of lax labor laws."

"Don't be ridiculous," I replied.

"At the least," he accused, "you'll hire cheaper minorities, and not give them any benefits."

"That may be true," I admitted, "but that's only because you're a white male who has been with the column a long time, so all new hires will be done at entry level salaries and minimal benefits. That's the American way! Besides, it'll only be just long enough to get women and other minorities as foils into my column, then I'll lay *them* off."

"What woman is going to be as big a schlemiel as I am?" a tearful Marshbaum asked. "I'm the best there is! No one can act dumber than me!"

I acknowledged it would be exceedingly difficult to find anyone like him. "You'll be able to get unemployment," I said. "Maybe even food stamps."

"I want a job," he protested. "The one I've held for years. Every time you were stumped for ideas, I did something stupid. Made you look like a genius. Stupidity is my life!"

"Marshbaum," I said, "don't you understand? If I use you much more, I'll probably be caught in some poll that academics conduct when they have nothing better to do with their time."

"But you use women," he said, his voice straining. "What about Susie Sweetwater, your shmuck of a TV news reader? What about—"

"Marshbaum," I said compassionately, "you're a victim of your own ability."

"What about Mike Royko?" Marshbaum protested. "You think he's gonna lay off Slats Grobnik? What about Art Buchwald? He's not going to give his foils their notice just because they're white men?"

"I believe they were grandfathered in," I said.

"Breslin!" Marshbaum thundered. "No one's telling Jimmy Breslin to get rid of Fat Thomas and Marvin the Torch!"

"Marshbaum," I said, "it'll only be a few weeks. A decade or two at the outside."

Reluctantly, he packed up his things and shuffled out the office. The last time I saw Marshbaum, he was reading a Spanish language book about trans-sexual surgery.

AUGUST 1993

Patriotic Unemployment

I was going into the Unemployment Office to pick up a copy of the monthly fiction from the Bureau of Labor Statistics when Robinson was coming out, smiling and holding what appeared to be an unemployment check. Robinson? The same Robinson who had worked more than 30 years in one job? Never late to work, never sick. Never promoted or demoted. Never stole; never argued; never complained. Loyal as a puppy.

"Robinson?" I asked cautiously.

"Hey, Man, good to see you. You still working?"

"Of course I'm still working. There's news to gather, stories to—"

I didn't have a chance to finish. He grabbed my arm, looked around, then dragged me into a corner away from traffic and peering eyes. I figured he had a choice piece of whistle-blowing.

"Don't say anything," he cautioned me. "I'll get you out of this."

"Out of *what*?" I asked nervously.

"Out of employment. Quit now before they get you."

"*Who'll* get me?"

"The feds."

"The *FBI* is after me?"

"FBI. CIA. VFW. Save yourself. Quit now and turn yourself in for unemployment compensation."

"You're a nut!" I proclaimed.

"Patriot!" he proudly corrected me. "I'm defending God, Mother, apple pie, the flag—remember never *ever* burn our Stars and Stripes—and Country. If it weren't for me, the country would collapse."

"Being unemployed is patriotic?" I asked suspiciously.

"Ever since some of the Administration's advisors claimed that a lower unemployment rate led to inflation, and that continued employment could lead to the collapse of the money market."

Perhaps some McFood he ate fried some of his few remaining McBrain cells. Robinson cautiously checked for feds, then took a crumpled news article from his back pocket, thrust it at me, and ordered me to read it. It seems that some red-taped administrators figured out that when unemployment is low, employers have to pay more to the workers since there are fewer of them available. The longer a worker stayed on the job, the higher the salary would be, and the higher salary would lead the employer to raise rates and prices to the consumers who would then demand more wages from their own employers. Conversely, a higher unemployment rate led to lower wages since employers could pick and choose who they wanted; this would then result in lower prices on consumer goods.

"You're a journalist!" Robinson proclaimed. "You're already suspect as being unpatriotic. Save yourself. Quit now."

"But, Robinson, I *like* my job."

"Think you'd like a few years in Leavenworth? Treason isn't taken lightly in this country.

"But I'm still earning a salary, and you're unemployed."

"And a *patriot*!" he reminded me.

"That may be true," I agreed, "but you're still only earning unemployment—"

"—welfare, food stamps, and AFDC!"

"AFDC?! Your wife left you six years ago. Both your children are out of college. How did you get Aid to Families with Dependent Children?"

"I adopted two Koreans. It was a patriotic responsibility. Did I ever tell you about Vocational Rehabilitation? Greatest thing. They pay me to learn a new job. All I have to do is go to school a couple of hours a day and collect a pay check."

"Well, at least that seems like something constructive from all this." I figured since there were mounting shortages of welders, nurses, and plumbers, he'd be enrolled in one of those programs.

"Journalism," he said matter-of-factly.

"Journalism!" I cried out. "Reporters are some of the most underpaid, overworked people in the country. When they can get jobs, that is. Haven't you read the numbers? Even with poor pay and benefits, thousands are trying to get into the field, but there aren't openings!"

"Exactly!" said Robinson, smiling. "Exactly."

AUGUST 1992

An Unequal Competition

And so there they were. The two of them. Locked into a professional struggle that neither of them knew about. He was a journalist. She was a young college graduate. Both wished to become teachers.

But she had one thing he didn't have. A state credential. She had taken almost two years of education courses. He hadn't.

Normally, he would have been one of the first to be weeded out, but somehow the clerks didn't notice he lacked the state credential. So there he sat. Facing the deputy superintendent of instruction, a man who had state credentials. An administrator who told him that without a general secondary credential the district wouldn't hire him.

"You need several courses. History of education. The philosophy of education. Educational methods. About fifteen in all, plus student teaching," he said.

"But I want to teach writing, not education," he replied.

"I understand that," said the deputy superintendent, "but we require that every teacher have a mastery of the concepts of education."

"I have a master's in journalism and professional experience," said the applicant, pointing out in one long breath-drenched sentence that in addition to having written for newspapers, magazines, and even television, he had a number of awards. "Isn't that enough?"

"I'm sorry, it isn't enough."

He explained he had taught creative writing part-time at a state university for three years, then pulled his relatively-high student evaluations from a folder. The deputy superintendent quickly scanned them, then put them aside.

"That doesn't eliminate the requirements. You can't teach in public schools until you learn *how* to teach."

"But I taught!" the writer argued.

"But you haven't taken education courses," said a most petulant deputy superintendent.

"I took two courses," said the writer, "and didn't see anything challenging in either of them." A little steamed, he went so far as to declare that most education courses seemed to be wastes. This insolence, of course, upset the deputy superintendent whose mission now was just to get rid of this pest.

"I would like to hire you, but—"

"But I don't have a credential! What about the other applicants? Do they have my experience? My writing ability? Or, do they only have your license?"

"The state requires," said the deputy superintendent, his attitude bound by a set of conventional regulations, "that a person have a number of courses in education. As I mentioned earlier, it's important to know *how* to work with students of this age group."

"Doesn't knowledge or ability count?" the writer demanded. But before the deputy superintendent could answer, the writer continued, "In one of your district schools, a guy with a psych degree and no media experience is advising the school newspaper."

The deputy superintendent coughed, then answered, "In that situation I believe he was already on the faculty, but due to a reduction in enrollment in psychology courses, the district either had to lay him off or find him another position."

"So you threw him into journalism."

"We didn't *throw* him anywhere!" snapped the deputy superintendent. "His credentials include teaching English, and he expressed an interest in teaching journalism."

"I have an interest in nuclear physics," the writer said sarcastically. "Of course, I don't know anything about it, but I'm interested in teaching it. I'll even take courses in education so I'll know nuclear physics better. And judging from what I've seen of your students, I could probably teach nuclear physics better than some of your English teachers are teaching writing!"

There was nothing anyone could have done. And so the young lady, with a degree in English who took all the education courses and completed student teaching but never wrote anything that was published, earned a position teaching creative writing and advising the school newspaper and literary magazine.

Occasionally, she mentioned something she learned from a class she took in college. But she never knew that the competition for her job had been her former teacher.

FEBRUARY 1994

An Hour a Day

It's dark and lonely at 3 a.m. on the 51 miles of roads between Bloomsburg and McAdoo in rural northeastern Pennsylvania. But, Bloomsburg is home, and McAdoo is where the Consolidated Cigar Corp. is, and that's where Jack W. Smith works. From 3:45 a.m. to 3:45 p.m. Almost every day, Mondays through Fridays, and occasionally Saturdays.

He says he'd move closer to his job, but he's hoping he won't have to work there much longer. He would like something better, but it's a job, and in this economy it's the only thing he can get. The job calls for him to take 300-pound rolls of paper and slit them into sizes to wrap tobacco. It was tough at first. His back and arms ached, and he suffered innumerable cuts. It's not easy when you're 54 years old and have never worked so hard in your life at physical labor.

"I'm one of the hardest working people they have," Smith says proudly, pauses a moment, then says, with equal pride, he's "just so happy to have a job." For most his adult life, he was employed. For four years, he was a cook in the Navy's submarine service. After that, he spent a year selling men's clothing. Then, in 1960, he got a job as a reporter on his hometown newspaper, the *Berwick (Pa.) Enterprise*, an 11,000 circulation afternoon daily that would eventually be merged into a sister publication in the early 1980s.

Jack Smith had never taken a journalism course, hadn't even gone to college, but the editor was willing to try an enthusiastic cub who said he learned quickly and could "find all the news." During the next 12 years, he proved his words, covering everything from PTA meetings to the police beat, digging out stories that reporters with years of experience and two college degrees couldn't find. And when his editor retired, Jack Smith was the natural replacement, a job he held for six years.

But, after 18 years as a journalist, he was 40 years old, not making a lot of money and had minimal benefits. He also had a wife, Gail, and three sons. It was time for a new job, one that gave him the freedom he never had. Taking a second mortgage on his house and everything in savings, he and Gail opened a gift store in 1979. In the next few years, they made the business one of the more successful ones in nearby Bloomsburg. So, he did what many businessmen did—he expanded, opening stores in Berwick, about 10 miles away, and in Williamsport, about 30 miles away. It was 1987, the end of the prosperity phase of the Reagan-Bush era and the beginning of the Recession. In 1989, he was forced to declare bankruptcy. He lost his house, most of his possessions and, more important, his self-respect. He was 52 years old.

He says he begged his former publisher for a job, but was curtly told there wasn't anything available. Not as a copyboy; not even as a janitor. He took state civil service tests, scoring in the 90s, but was never hired, even for entry-level public information jobs, something even 21-year-olds with a fraction of Smith's knowledge and experience could do. He was the highest scorer in a five-county area for a state social service job, but was never hired. He applied for other jobs, but says in most cases he "never got past the receptionists."

Maybe one of his poker buddies could help him out, he thought. After all, they had been friends for several years. But the one who was an executive at a lingerie factory and the one who was partial owner of a carpet mill said there wasn't anything. And the longer Smith was unemployed, the farther they strayed from friendship, and Smith never knew why.

He had no job, his savings was gone, and he had no health insurance. Not many unemployed people can afford almost $5,000 a year for health insurance.

For a few months in 1990, he worked for the Census Bureau, eventually becoming a field supervisor in a five-county area. He got the Bureau's highest evaluations, but the job ended, and he was again unemployed. He was even desperate enough to consider several minimum wage jobs, knowing that an income of $8,840 a year wasn't enough to support one person, let alone a family. Alas, he didn't even get the minimum wage jobs.

Once, he even applied for a job cleaning wards in the VA Hospital in Wilkes-Barre, but never even got an interview. But that's how he got the job at the cigar factory. The owner of the plant, a friend who thought it would be demeaning to offer a job as a laborer to someone who had been a Navy vet, newspaper editor, and successful businessman, told Smith's wife, "If he's going to wash walls, have him come see me." So, Jack Smith became the oldest slitter in the company. In six months, he lost 30 pounds as his body wearied under the physical demands of his new job. Yet, he stayed with it, taking all the overtime he could. "When I don't take the overtime," he says, "my check doesn't stretch as it should." It stretches even less now since his rent was increased 30 percent.

"The landlord said she had trouble meeting expenses," Smith says. This month, he starts repaying a college loan. He had begun taking college classes, one or two at a time, shortly before his business failed. He had earned 47 credits when he ran out of money.

Gail has taken part-time jobs and stood by him, although she cries a lot. "My biggest joy is my family," says Smith, himself shedding a small tear when he says "We've become more cohesive the past few years. We're hanging tough together."

It's been almost two years now since Jack Smith was first hired as a slitter. It's been four years since his business failed, and society remembers his failure, not his success. He keeps applying for jobs. He keeps getting rejected. He's optimistic about the job he's doing, about his own life, but has lost some of his faith in the country that says it values opportunity, age, experience, and ability. It is something he thinks about five mornings a week as he drives an hour in the dark to get to work as a laborer.

MARCH 1993

In 1994, Jack Smith began working as a bartender in a Berwick restaurant. Almost two years later, he and his wife, Gail, opened a co-op crafts store in a converted church.

'We're Management; We Don't Have to Tell You Anything'

Shirley Collins sat in a nondescript office, facing the editor of the *Globe-Times* of Bethlehem, Pennsylvania. Nearby sat an outside consultant and an attorney from Chicago. We're restructuring the newspaper, said the editor, reading from a prepared script; your work is unsatisfactory, he claimed. You're being terminated, he emphasized.

In the seven years she had been at the newspaper as feature news editor, Shirley Collins had never received a formal evaluation from any of her supervisors, although she did receive several awards, the most recent being second-place for lifestyle sections in the annual Keystone awards contest of the Pennsylvania Newspaper Publishers Association. But, the well-rehearsed script the company provided its participants April 5, 1988, didn't allow the editor to give individual evaluations the day of her firing. Nor was it likely they would. Two months earlier, the newspaper fired Mary Wagner, a 21-year employee who was one of the better police reporters in America. When the staff, shaken by her firing, asked the editor why Wagner was fired, he coldly replied to two of their representatives, "We're management. We don't have to tell you anything."

On the day she was fired, Shirley Collins sat there, and listened to an editor claim that her work was "unsatisfactory." We have an agreement, the editor read, explaining that if she signed the eight-page agreement and resigned rather than be fired, she would receive "extra benefits," including about one week of severance pay for each year worked, continuation of health and life insurance benefits for a limited time and, if eligible, be able to collect all pension benefits due.

Of course, there were a few "trade-offs" in order for Collins to receive all these "extra benefits"—she would have to agree to give up many of her rights. For one year, she would be forbidden, without *Globe-Times* consent, to associate in any way with any newspaper or magazine publishing company or any radio or television station that broadcast news programs within 25 miles of the *Globe-Times*; for two years, she would not be allowed to associate with the Times Mirror Co., publisher of the competing Allentown *Morning Call*, or Thomson Newspapers, Inc., publisher of the nearby *Easton Express*. Not only couldn't she work for those organizations, she couldn't even deliver newspapers for them or rent an apartment if the owner was one of the forbidden companies.

Among other rights she would have lost had she signed the agreement, she would have had to agree to never disclose the contents of the termination agreement, except to immediate family and her attorney; "cooperate with and assist The Globe-Times in any investigations, proceedings, or actions" relating to her employment or "to any matter in which I was involved or of which I had knowledge of while an employee of, or a consultant for, The Globe-Times"; and pay all costs and expenses incurred by the company should it bring any charges against her for violation of the agreement, or should she bring any charges against the company "relative [to] any such action, proceeding, claim or charge."

And then there was Paragraph 6 which deleted her statutory and Constitutional rights, including the right to bring charges against the company for violations of Title VII of the Civil Rights Act, the Age Discrimination in Employment Act of 1967, the Fair Labor Standards Act, the National Labor Relations Act, and the Pennsylvania Human Relations Act.

Shirley Collins refused to sign the agreement. She was again urged to consider the many "extra benefits" she would get from signing, and was told that Management was sorry that she chose not to take advantage of those fine benefits. She didn't know why she was being fired. She was a 55-year-old woman, with a master's degree in library science, and another master's degree in journalism from Columbia University, which she had earned shortly before she was hired at the *Globe-Times*—and she was unemployed.

By the end of the day, more than two dozen employees would be fired, including more than 40 percent of the editorial staff. By the end of the year, there would be several resignations, and more than 50 would have been fired; about two-thirds of the editorial staff would have been fired or left voluntarily as Management shuffled personnel, trying to find what it believed was the right combination to bring the newspaper out of a financial tailspin brought about when circulation plummeted from 39,572 in March 1981 to 21,702 in March 1988.

In an industry of one-newspaper towns, the *Globe-Times*, at one point one of the nation's better smaller newspapers, should have had its own comfortable monopoly. But in the 1980s, like most newspapers, it hadn't effectively planned how to compete with a regional daily, alternative sources of advertising, a recession, a population shift to morning newspapers, a reduction in the reading habits of most Americans, and a demographic time bomb that exploded when the advertising industry increased its emphasis upon quality numbers rather than mere quantity.

The *Globe-Times* solution was a massive restructuring, focusing upon image and a misguided effort to promote what it thought was "community journalism" at the expense of the hard news stories readers needed. But it became apparent with the one-day purge. Many were in their 40s and 50s, had worked at the newspaper two or more decades, and were arguably among the better workers in the industry. The purge left the newsroom decimated, stories uncovered; "happy news" filling the front page, and the readers wondering why they should even subscribe anymore.

Most of the staff who remained were uncertain of the direction the newspaper was taking—or even if they would have jobs the next week. By the end of 1989, even the editor who had fired Mary Wagner, Shirley Collins, and more than a dozen other reporters was gone.

Pennsylvania law, like that of most states, gives private-sector employers in non-union operations the right to hire and fire "at will." The "employment at will doctrine," with very few exceptions, essentially states that private sector non-unionized employees are at the mercy of the employer. The concept is that the management may fire an employee "for good cause, for no cause, or even for cause morally wrong . . ." [*Payne v. Western & Atlantic RR*],

Tennessee, 1884.] The United States is the only major industrial democracy that does not have federal legislation prohibiting unfair discharge. About twenty-two million Americans work under collective bargaining agreements; sixty million don't.

Most employees believe employers have a right to hire and fire "at will," not knowing that employment in much of the public sector and on unionized newspapers, magazines, radio, or television stations and networks, gives the worker innumerable benefits and rights, including the right to be fired only for "just cause."

If the story of the *Globe-Times* had applied only to one newspaper, it could easily be dismissed as unique in American journalism. Journalists and the public could look over the events, argue about them, perhaps sympathize with the workers who were fired or with management that says it needed to make changes to regain financial stability; perhaps many might even think that some of the employees deserved to be fired; perhaps some would think that management was morally and ethically wrong in how and why it fired many of the staff.

However, the issues aren't of personalities or events, or even in management's decisions that led to the crisis or its efforts to recover. The story is about labor-management relationships not only at the *Globe-Times* but in all private industry. The issues are universal in American labor, but more so when members of the editorial staff must carry the burden of loyalty to their profession and the community against loyalty to the corporation that pays their salaries, knowing that they could be reprimanded, demoted, or fired at any time for any reason.

Had there been a union at the *Globe-Times*, the employees would have been able to take advantage of well-established multistep grievance procedures detailed in the collective bargaining agreement, including the right to go to binding arbitration in the case of termination or arbitrary job changes. In the United States, about half the union-protected workers whose cases go to binding arbitration are restored to their jobs, often with back pay. The collective bargaining agreement would also have detailed specific agreements with respect to promotion, termination, restructuring, or a layoff. Further, the employees would have been protected by state and federal labor codes.

Without union representation, the workers had yielded equity, for management has the money and resources to bring in consultants and attorneys to keep former employees, most of whom can't afford $150-$200 an hour lawyer fees, from mounting a significant legal challenge.

But there is still a "stigma" within the journalism profession about being a union member. "It's OK for blue-collar workers," goes the refrain, "but we're *professionals*." Perhaps one day, the *professionals* will realize that in the United States, the division by traditional class—lower, middle, upper—is no longer viable, and there truly has been a division along lines of the underclass, the workers, and the management; journalists must again recognize that they have far more in common with the coal miner than they have with the chief executive officer. The tragedy that is partially shared by the *Globe-Times* is the failure of society to adequately protect all its workers.

On November 4, 1991, the Globe-Times *was merged into the competing* Easton Express. *For a full story of the* Globe-Times *see* Death of an American Newspaper, *by Walter M. Brasch, published by Associated University Presses, 1996.*
APRIL 1989

Withdrawing From an O.J. Overdose

The abrupt end of "Days of Our O.J." caught Americans by surprise, leaving them confused, unable to function, and whining that three hours were long enough for a football game but not acceptable for a jury's deliberation.

Also in withdrawal are the media. More than 1,000 reporters, cramped into the parking lot known as Camp O.J., were credentialed to cover the trial. Thousands more had some part in O.J. coverage; CNN alone devoted more than 600 hours, and 425 reporters and commentators to the trial; the other networks devoted more time to the case than any other story, including the Oklahoma bombing, floods in the Southeast, and the ethnic slaughter in Bosnia. Unable to quit cold turkey, the media are desperately milking the story of The Juice. Within minutes of the verdict, it seemed as if every reporter in L.A. bought ladders, just so they could peer into O.J.'s Brentwood yard and continue to give the people their daily dose of O.J. in the form of all the gossip that's fit to be smeared.

Newspaper editors, who have been blaming the "newsprint crisis" for increased subscription costs and fewer newspaper pages, laid off reporters last year, but always made sure there was sufficient room in the paper for O.J. coverage. After the verdict, they have met the news hole by publishing summaries of every hour of testimony, and what the reporters on ladders are really seeing in

Brentwood, with interminable sidebar opinions of everyone who ever watched a football game or drove a rental car and, therefore, absolutely knew whether or not O.J. was innocent.

But the most tragic consequence of the ending of the trial is that most of the nation's lawyers are now unemployed. If they weren't working for the Prosecution or the Defense, they were commentators for the TV networks.

Whenever TV devotes more than 90 seconds to any story, it hires "expert" commentators. During the Gulf War, the networks provided full employment to retired generals. This time, it's lawyers. The commentators, like most TV anchors, tend to be White, college-educated, TV-quality clean-shaven handsome and makeup beautiful, and upper middle-class or upper-class. Thus, a typical day's news coverage included lengthy snippets from the trial, reporters' comments, reporters talking to each other for their reactions, and reporters talking to lawyers who believed it was their sacred duty to uncover every mole on the face of the trial, yet missed the greater body of criminal justice issues.

The media's failure to recognize and hire others who would have knowledge about crime in America is symptomatic of the media's fetish with what they believe are the "experts." It's no different when two people come into a newsroom—a well-groomed Suit with a press release about some owner's nephew getting a promotion, and a bag lady with information about police abuse of the elderly. Does anyone doubt which one the reporters will talk to first—assuming Security hasn't thrown the bag lady out already?

Also missing while the media were covering the trial of the century was coverage of thousands of murders that happened, and that prosecutors often plea-bargained justice in order to clear the overburdened court calendars, while failing to even acknowledge the victim's families. Also overlooked is that most defendants can't afford "dream team" representation, and receive public defenders who are only a few years out of law school and whose budgets barely cover legal pads. Would it have been too much of a burden for the media to devote a few hours to probing the quality of justice in America when the defendant isn't a millionaire celebrity?

Should the media decide to spend time covering the greater social issues instead of trying to justify themselves by coming up with 100 reasons why the O.J. trial changed civilization, it might look into doing extended coverage of the issues of spousal abuse, the homeless, the health care crisis, street crime, the burgeoning right-wing survivalist and militia movements, teenage alcohol and drug abuse, the economy, America's political process, and the consequences of catastrophic fires, floods, and hurricanes.

The media also overlooked society's achievements. It might be nice if the media gave as much coverage to the lives of volunteers for the nation's social service agencies as they did for the exploits of $400 an hour lawyers.

In response, the media titans say they are giving the people what they want, that they are merely feeding the public's frenzy for the sensational. But by devoting the top five minutes of the 6:30 p.m. newscast to O.J., and only 90 seconds to floods in the Florida panhandle, or giving O.J. a 6-column newspaper banner and the Bosnian holocaust a 2-column headline beneath the fold, the media have told us what they think is important—or what they think the public thinks is important—and we somehow believe them because it's easier for them—and us—to ignore the greater issues. Since the people aren't complaining, the media assume that's what the people want.

When the media increase their coverage of all society instead of one abhoration, then maybe they can devote a few minutes or column inches to the celebrity crime scene.

OCTOBER 1995